Solid Waste
and Emergency Response
(5203P)

EPA 542-R-09-007
September 2010
www.clu-in.org/POPs

# Reference Guide to Non-combustion Technologies for Remediation of Persistent Organic Pollutants in Soil, Second Edition – 2010

**Internet Address (URL)** http://www.epa.gov

**CONTENTS**

**Appendix**

A      Chemical Structures, Uses and Effects of POPs listed under the Stockholm Convention and LRTAP

B      Fact Sheet on Anaerobic Bioremediation Using Blood Meal for the Treatment of Toxaphene in Soil and Sediment

C      Fact Sheet on Bioremediation Using DARAMEND® for Treatment of POPs in Soils and Sediments

D      Fact Sheet on In Situ Thermal Desorption for Treatment of POPs in Soils and Sediments

E      Additional Technologies Identified but Not Commercially Available

## LIST OF TABLES

## ACRONYMS AND ABBREVIATIONS

| | |
|---|---|
| ART | Adventus Remediation Technologies, Inc. |
| ATSDR | Agency for Toxic Substances and Disease Registry |
| BCD | Base-catalyzed decomposition |
| BDE | Bromodiphenyl ether |
| BHC | Hexachlorobenzene |
| BFR | Brominated flame retardants |
| CaO | Calcium oxide |
| CB | Chlorobenzene |
| CCMS | Committee on the Challenges of Modern Society |
| CD | Catalytic dechlorination |
| CFC | Chlorofluorocarbon |
| CHD | Catalytic hydrodechlorination |
| CLU-IN | Clean-Up Information |
| COP | Conference of the Parties |
| cy | Cubic yard |
| DDD | Dichlorodiphenyldichloroethane |
| DDE | Dichlorodiphenyldichloroethylene |
| DDT | Dichlorodiphenyltrichloroethane |
| DDX | Total Dichlorodiphenyldichloroethane, Dichlorodiphenyldichloroethylene, and Dichlorodiphenyltrichloroethane |
| Dioxins | Polychlorinated dibenzo-p-dioxins |
| DNT | Di-nitro toluene |
| DRE | Destruction and removal efficiency |
| EDL | Environmental Decontamination Ltd. |
| US EPA | U.S. Environmental Protection Agency |
| ERT | Environmental Response Team |
| FAO | Food and Agricultural Organization |
| FRTR | Federal Remediation Technologies Roundtable |
| ft | Foot |
| Furans | Polychlorinated dibenzo-p-furans |
| GEF | Global Environmental Facility |
| GPCR | Gas-phase chemical reduction |
| gpd | Gallon per day |
| GRB | Gila River Boundary |
| GRIC | Gila River Indian Community |
| HBCDD | Hexabromocyclododecane |
| HCB | Hexachlorobenzene |
| HCBD | Hexachlorobutadiene |
| HCH | Hexachlorocyclohexane |
| HEPA | High-efficiency particulate air |
| HMX | High melting explosive, octahydro-1,3,5,7-tetranitro-1,3,5,7 tetrazocine |
| HTTD | High temperature thermal desorption |
| ICV | In Container Vitrification |
| IHPA | International HCH and Pesticides Association |
| ISTD | In situ thermal desorption |
| ISV | In situ vitrification |
| JESCO | Japan Environmental Safety Corporation |
| kg | Kilogram |
| LRTAP | Long-Range Transboundary Air Pollution |

| | |
|---|---|
| LTR | Liquid tank reactor |
| LTTD | Low temperature thermal desorption |
| M | Meter |
| MCD | Mechanochemical dehalogenation |
| MDL | Method detection limit |
| MTBE | Methyl tert-butyl ether |
| µg/kg | Microgram per kilogram |
| mg/kg | Milligram per kilogram |
| mm | Millimeter |
| ng-TEQ/g | Nanogram Toxic Equivalent of Dioxins per gram |
| NA | Not available |
| NAPL | Nonaqueous-phase liquid |
| NATO | North Atlantic Treaty Organization |
| ND | Not detected (indicating concentration below detection limit) |
| NIP | National Implementation Plan |
| OSRTI | Office of Superfund Remediation and Technology Innovation |
| Pa | Pascal |
| PACT | Plasma Arc Centrifugal Treatment |
| PAH | Polycyclic aromatic hydrocarbons |
| PCB | Polychlorinated biphenyl |
| PCD | Photochemical dechlorination |
| PCN | Polychlorinated naphthalenes |
| PCNB | Pentachloronitrobenzene |
| PCP | Pentachlorophenol |
| PCS | Plasma Converter System |
| PFOS | Perfluorooctane |
| pg-TEQ/g | Picogram Toxic Equivalent of Dioxins per gram |
| POP | Persistent organic pollutant |
| POPRC | Persistent Organic Pollutants Review Committee |
| ppb | Part per billion |
| ppm | Part per million |
| ppt | Part per trillion |
| PVC | Polyvinyl chloride |
| RCRA | Resource Conservation and Recovery Act |
| REACHIT | Remediation and Characterization Innovative Technologies |
| rpm | Revolutions per minute |
| SCCP | Short-chained chlorinated paraffins |
| SCWO | Supercritical water oxidation |
| SITE | Superfund Innovative Technology Evaluation |
| SOx | Sulfur oxide |
| SP | Sodium Powder Dispersion Dechlorination Process |
| SPHTD | Self-propagating high-temperature dehalogenation |
| SPV | Subsurface Planar Vitrification |
| SR | Sodium reduction |
| STAP | Science and Technology Advisory Panel |
| SVOC | Semivolatile organic compound |
| t-BuOK | Potassium tert-butoxide |
| TCDD | Tetrachlorodibenzodioxin |
| TCLP | Toxicity Characteristic Leaching Procedure |
| TSCA | Toxic Substance Control Act |
| TNT | Trinitrotoluene |

| | |
|---|---|
| UNECE | United Nations Economic Commission for Europe |
| UNEP | United Nations Environment Programme |
| UNR | University of Nevada at Reno |
| USD | United States Dollar |
| VOC | Volatile organic compound |
| WCS | Wasatch Chemical Superfund |

# NOTICE AND DISCLAIMER

This report compiles information about non-combustion technologies for remediation of persistent organic pollutants, including technology applications at both domestic and international sites, but is not a comprehensive review of all the current non-combustion technologies or vendors. This report also does not provide guidance regarding the selection of a specific technology or vendor. Use or mention of trade names or commercial products does not constitute endorsement or recommendation for use.

This report has undergone U.S. Environmental Protection Agency (US EPA) and external review by experts in the field. However, information in this report is derived from many references (including personal communications with experts in the field), some of which have not been peer-reviewed.

This report was prepared by the US EPA Office of Superfund Remediation and Technology Innovation (OSRTI), with support provided under Contract Numbers 68-W-02-034 and EP-W-07-078. For further information about this report, please contact Michele Mahoney at US EPA's Office of Superfund Remediation and Technology Innovation, at (703) 603-9057, or by e-mail at mahoney.michele@epa.gov.

A PDF version of "Reference Guide to Non-combustion Technologies for Remediation of Persistent Organic Pollutants in Soil, Second Edition-2010" is available for viewing or downloading at the Hazardous Waste Cleanup Information System website at *http://www.clu-in.org/POPs*. A limited number of printed copies of the report are available free of charge and may be ordered via the website, by mail, or by fax from the following source:

US EPA/National Service Center for Environmental Publications
P.O. Box 42419 USEPA
Cincinnati, OH 45242-2419
Telephone: 800-490-9198
Fax: 301-604-3408
Website: www.epa.gov/nscep

# EXECUTIVE SUMMARY

This report is the second edition of the U.S. Environmental Protection Agency's (US EPA's) 2005 report and provides a high level summary of information on the applicability of existing and emerging non-combustion technologies for the remediation of persistent organic pollutants (POPs) in soil. POPs are chemicals that are demonstrated to be toxic, persist in the environment for long periods of time, and bioaccumulate and biomagnify as they move through the food chain. POPs are linked to adverse effects on humans and animals, such as cancer, damage to the nervous system, reproductive disorders, and disruption of the immune system. In addition, restrictions and bans on the use of POPs have resulted in a significant number of unusable stockpiles of POP-containing materials, largely located outside the United States (US). Deterioration of storage facilities used for the stockpiles, improper storage practices, and past production and use of POPs also have resulted in contamination of soils around the world. Since the publication of this report in 2005, nine (9) additional chemicals have been listed in the Stockholm Convention; this brings the total number of chemicals currently listed as POPs under the Stockholm Convention to twenty-one (21)[1]. In addition, three (3) POPs are currently under consideration.

Historically, POP-contaminated soil has been widely treated by combustion systems using high temperature incineration to destroy the contaminants. Incineration is widely used because high-temperature incinerators can address large volumes of contaminated material and can treat most POPs contaminants. Modern incinerators operating with highly controlled combustion environments can achieve a high destruction and removal efficiency (DRE) for POP contaminants. In the US, DREs as high as 99.9999% are achievable for incinerators treating non-liquid polychlorinated biphenyl compounds (PCBs). The US EPA has approved the use of incinerators to treat PCB-contaminated material with PCB concentrations greater than 50 parts per million (ppm). However, US EPA requires that incinerators meet stringent operating conditions. For example, incinerators treating liquids contaminated with PCBs are required to meet either (1) a 2-second residence time for the liquid waste at a temperature of $1200\,^{0}C$ and with 3 percent excess oxygen in the stack gases or (2) a 1.5-second residence time at $1200\,^{0}C$, with 2 percent excess oxygen.

Though incinerators can be used to treat POPs, they have several technology limitations, which are addressed in the body of this report. Also, many interested parties have expressed concern about potential environmental and health effects associated with this type of treatment technology (Ref. 8). The combustion of POPs can create by-products such as polychlorinated dibenzo-p-dioxins (i.e., dioxins) and polychlorinated dibenzo-p-furans (i.e., furans) – two known human carcinogens. Due to concerns about their safety, incinerators also can face negative public opinion and attract public opposition. However, because alternative treatment approaches have been limited to date, incineration continues to be most commonly used technology for the treatment of POPs (including in developing countries). Additional information about incineration and other combustion technologies can be obtained from the US EPA's Federal Remediation Technologies Roundtable (FRTR) website (http://www.frtr.gov/matrix2/section3/3_6.html).

As a result of widespread interest in alternate technologies, numerous international organizations have developed reports that identify and discuss non-combustion technologies for POPs, including:

- Evaluation of Demonstrated and Emerging Remedial Action Technologies for the Treatment of Contaminated Land and Groundwater (Phase III), 2002. IHPA.

---

[1] http://chm.pops.int/Convention/ThePOPs/tabid/673/language/en-US/Default.aspx

http://www.ihpa.info/resources/library

- Review of Emerging, Innovative Technologies for the Destruction and Decontamination of POPs and the Identification of Promising Technologies for Use in Developing Countries, 2004. UNEP. http://www.basel.int/techmatters/review_pop_feb04.pdf
- Destruction and Decontamination Technologies for PCBs and Other POPs Wastes (Part III Annexes) A Training Manual for Hazardous Waste Project Managers, Volume C, 2005. Basel Convention. http://www.basel.int/meetings/sbc/workdoc/TM-A.pdf
- Non-Combustion Technologies for POPs Destruction – Review and Evaluation, 2007. International Centre for Science and High Technology. http://www.ics.trieste.it
- Updated general technical guidelines for the environmentally sound management of wastes consisting of, containing or contaminated with persistent organic pollutants (POPs). 2007. Basel Convention. http://www.basel.int/pub/techguid/tg-PCBs.pdf
- Disposal Technology Options Study – review and update of technology Annex C: Review and Update of Technology, 2008. Africa Stockpiles. http://www.africastockpiles.net

Some of the technologies discussed in these documents have progressed from the development stage to a commercial stage; other technologies presented as commercial stage are no longer being developed. In addition, promising destruction technologies for POPs continue to be developed. This Reference Guide to Non-combustion Technologies for Remediation of POPs in Soil is intended to summarize and update the First Edition prepared by US EPA in 2005, and build on these more recent studies. Updated information for this document was obtained primarily by (1) reviewing various websites and documents, (2) contacting technology vendors and experts in the field, and (3) working closely with the International Hexachlorocyclohexane (HCH) and Pesticides Association, IHPA (John Vijgen), which has published several factsheets that are used as references for this report.

This Second Edition Report also provides new performance data of the non-combustion technologies. Tables 3-1 and 3-2 summarize full-scale and pilot/bench-scale technologies and provide information on waste strength treated, ex situ or in situ treatment applicability, contaminants treated, available cost information, pretreatment requirements, power requirements, configuration needs, and links to individual fact sheets. Fact sheets prepared by US EPA are provided as appendices to this report. Additional fact sheets for the various technologies are available through the IHPA website. Technologies identified in the first edition (2005) of this report that are not currently commercially available are described in Appendix E. This document is not intended as a roadmap for technology selection; however it is intended to present the current state of knowledge for non-combustion technologies for treatment of POPs in soils.

## 1.0    INTRODUCTION

Persistent organic pollutants (POPs) are toxic chemicals that are chemically stable, do not easily degrade in the environment, and tend to bioaccumulate and biomagnify as they move through the food chain. Serious human health problems are associated with exposure to POPs, including cancer, neurological damage, birth defects, sterility, and immune system suppression. Restrictions and bans on the use of POPs chemicals have resulted in a significant number of unusable stockpiles of POP-containing materials internationally. In addition, deterioration of storage facilities used for the stockpiles, improper storage practices, and waste/releases associated with past production and use of POPs have resulted in contamination of soils around the world. The Programme on the Prevention and Disposal of Obsolete Pesticides by the Food and Agricultural Organization (FAO) of the United Nations is creating an inventory of obsolete pesticides stockpiled around the world. Information about pesticides inventories by country can be obtained from FAO of the United Nations at http://www.fao.org/ag/AGP/AGPP/Pesticid/Disposal/en/492 74/index.html. Because of their chemical stability, tendency to bioaccumulate, adverse health effects, and widespread distribution and presence, remediation technologies are needed to treat these pollutants.

> **FURTHER INFORMATION ABOUT THE LOCATION OF STOCKPILES IS AVALABLE AT**
> http://www.fao.org/ag/AGP/AGPP/P esticid/Disposal/en/49274/index.html

The international community has responded to the health concerns posed by these unusable stockpiles of POPs by developing various treaties and organizations to address POPs chemicals and waste. Under the Stockholm Convention on POPs (Stockholm Convention), which was adopted in 2001 and enacted in 2004, parties committed to reduce or eliminate the production, use, and release of the 12 POPs of greatest concern to the global community. The US is a signatory to the Stockholm Convention on POPs – but has not yet ratified the Convention. The initial list of 12 POPs was identified by the Intergovernmental Forum on Chemical Safety and the International Programme on Chemical Safety. Another treaty regulating POPs internationally is the Basel Convention on the Control of Transboundary Movements of Hazardous Wastes and their Disposal (Basel Convention). The Basel Convention was adopted on March 22, 1989 by the Conference of Plenipotentiaries convened at Basel, Switzerland and entered into force in 1992. In response to Stockholm Convention provisions requiring coordination with the Basel Convention on POPs waste issues, the Basel convention developed guidance on the environmentally sound management of POPs waste. In 2004, the Basel Convention invited signatories of the Stockholm Convention to consider its recommendations on environmentally sound management for POPs wastes (Refs. 69 and 70). The US is a signatory to the Basel Convention on POPs – but has not yet ratified the Convention.

> **FURTHER INFORMATION ABOUT THE STOCKHOLM CONVENTION ON POPs IS PROVIDED AT** http://chm.pops.int/

The Stockholm Convention's subsidiary body – the POPs Review Committee (POPRC[2]) – includes environmental experts that review proposals to add new chemicals to the Convention. The POPRC uses criteria set forth in the Convention[3] to review a chemical's characteristics as well as human health and environment effects. If the chemical meets the Convention's screening criteria in Annex D to the Convention, then the POPRC develops a risk profile for the chemical (according to Annex E of the Convention) and, if warranted, prepares risk management evaluation

---

[2] Additional information about POPRC meeting can be found at:
http://www.pops.int/documents/meetings/poprc/poprc.htm
[3] See Article 8 and Annexes D, E, and F of the Stockholm Convention.

(according to Annex F of the Convention). Upon completion of the risk management evaluation, the POPRC makes the recommendation to the Conference of the Parties (COP) whether or not to add the chemical to one (or more) of the Convention's Annexes (i.e., Annexes A, B, and/or

> **FURTHER INFORMATION ABOUT THE CONVENTION ON LONG-RANGE TRANSBOUNDRY AIR POLLUTION IS PROVIDED AT**
> http://www.unece.org/env/lrtap/welcome.html

C). In addition, part of the listing process is to aid in creating a plan to reduce the chemical from current and future environmental applications or uses (Ref. 61).

In October 2008, the POPRC held its fourth meeting (POPRC-4) and an outcome of that meeting was that it recommended to the May 2009 COP that nine (9) additional chemicals be added to the Stockholm Convention (Ref. 27). In October 2009[4] at POPRC-5 meeting, several chemicals underwent a review process by the Committee. However, no new chemicals were recommended by the POPRC to the COP for listing. Future COP and POPRC meetings in 2010 and 2011 will continue to review chemicals and possibly add new chemicals to the Stockholm Convention.

In addition to the (global) Stockholm Convention, the Convention on Long-Range Transboundary Air Pollution (LRTAP) is a regional international treaty that addresses environmental issues of the United Nations Economic Commission for Europe (UNECE) with a primary focus on air emissions. The US is a Party to the LRTAP Convention (i.e., the US has signed and ratified the LRTAP Convention). The LRTAP Convention has been extended by eight (8) Protocols that include specific requirements for countries to reduce air pollution including long-range air pollution. In 1998, the LRTAP Convention adopted a Protocol on POPs to regulate the production and use of 16 chemicals that were singled out according to agreed risk criteria. The US is a signatory to the LRTAP's Protocol on POPs – but has not yet ratified the Protocol and, therefore, not yet a Party to the LRTAP's Protocol on POPs. The LRTAP POPs Protocol originally listed the following 16 chemicals when it was adopted in 1998: aldrin, chlordane, chlordecone, dichlorodiphenyltrichloroethane (DDT), dieldrin, dioxins/furans, endrin, heptachlor, hexabromobiphenyl, hexachlorobenzene (HCB), lindane (i.e., gamma-HCH), mirex, polychlorinated biphenyls (PCB), polycyclic aromatic hydrocarbons (PAHs), and toxaphene. On December 18, 2009[5], the parties to the LRTAP POPs Protocol adopted amendments to the protocol to include seven (7) additional chemicals, which are: octabromodiphenyl ether, pentabromodiphenyl ether, pentachlorobenzene, perfluorooctane sulfonate (PFOS), short-chained chlorinated paraffins (SCCP), polychlorinated naphthalenes (PCN) and hexachlorobutadiene (HCBD). PCN and HCBD are the only two chemicals listed in the LRTAP's Protocol on POPs that are not already listed (or under review for listing) by the Stockholm Convention.

Table 1-1 lists all of the 26 chemicals identified under the Stockholm Convention on POPs and the LRTAP's Protocol on POPs - both currently listed and under review.

---

[4] The final report from the POPRC-5 meeting can be found at: http://chm.pops.int/
[5] The final report from the LRTAP's Executive Body meeting can be found at: http://www.unece.org/env/lrtap/ExecutiveBody/welcome.27.html

| Table 1-1. POPs Identified by the Stockholm Convention and Long-Range Transboundary Air Pollution Convention | | | |
|---|---|---|---|
| | Stockholm Convention | | Long-Range Transboundary Air Pollution Convention |
| POP | Currently Listed | Under Review (2009) | |
| **Pesticides** | | | |
| Aldrin | ✓ | | ✓ |
| **Alpha-hexachlorocyclohexane** | ✓ | | ✓ |
| **Beta-hexachlorocyclohexane** | ✓ | | ✓ |
| Chlordane | ✓ | | ✓ |
| **Chlordecone** | ✓ | | ✓ |
| Dichlorodiphenyltrichloroethane (DDT) | ✓ | | ✓ |
| Dieldrin | ✓ | | ✓ |
| Endosulfan | | ✓ | |
| Endrin | ✓ | | ✓ |
| Heptachlor | ✓ | | ✓ |
| Hexachlorobenzene (HCB) | ✓ | | ✓ |
| **Lindane** | ✓ | | ✓ |
| Mirex | ✓ | | ✓ |
| Toxaphene | ✓ | | ✓ |
| **Industrial Chemicals or By-Products** | | | |
| Dioxins | ✓ | | ✓ |
| Furans | ✓ | | ✓ |
| **Hexabromobiphenyl** | ✓ | | ✓ |
| Hexabromocyclododecane (HBCDD) | | ✓ | ✓ |
| Hexachlorobutadiene (HCBD) | | | ✓ |
| **Octabromodiphenyl ether** | ✓ | | ✓ |
| **Pentabromodiphenyl ether (penta-BDE)** | ✓ | | ✓ |
| **Pentachlorobenzene** | ✓ | | ✓ |
| **Perfluorooctane sulfonate (PFOS)** | ✓ | | ✓ |
| Polychlorinated biphenyls (PCB) | ✓ | | ✓ |
| Polychlorinated naphthalenes (PCN) | | | ✓ |
| Short-chained chlorinated paraffins (SCCP) | | ✓ | ✓ |

Sources: Refs. 27, 61and 68
Note: Nine additional chemicals that were recently listed by the Stockholm Convention in May 2009 are shown in bold.

Historically, POP-contaminated soil and stockpiles have been widely treated using combustion systems using high temperature incineration to destroy the contaminants. Incineration is widely used because high-temperature incinerators can address large volumes of contaminated material and can treat most contaminants. Modern incinerators operating with highly controlled combustion environments can achieve a high destruction and removal efficiency (DRE) for POP contaminants. In the US, DREs as high as 99.9999% are achievable for incinerators treating non-liquid PCBs. US EPA has approved the use of incinerators to treat PCB-contaminated material with PCB concentrations greater than 50 parts per million (ppm). The US EPA requires that incinerators meet stringent operating conditions. For example, incinerators treating liquids contaminated with PCBs are required to meet a either: (1) a 2-second residence time for the liquid waste at a temperature of 1200°C and with 3 percent excess oxygen in the stack gases or (2) a 1.5-second residence time at 1200°C, with 2 percent excess oxygen (40 C.F.R. § 761.70). Because of its capabilities, incineration is a viable option for the treatment of materials containing POPs.

There are several limitations in the reliance on incineration as the sole alternative for POPs waste treatment. For example, incinerators cannot destroy inorganic constituents (metals) in waste streams, and these maybe released in air emissions or retained in solid residues; therefore, waste containing POPs and certain metals may not be suitable for incineration in some cases. Some heavy metals (including lead, cadmium, mercury, and arsenic) may partially vaporize and leave the combustion unit of the incinerator with the flue gases; this can require additional off-gas treatment systems for removal of these gaseous combustion products. Incinerators treating waste streams contaminated with heavy metals can also produce a bottom ash with high concentrations of metals. These bottom ashes then require characterization to determine whether they are Resource Conservation and Recovery Act (RCRA) hazardous waste, may require stabilization, and must be disposed of appropriately. Also, combustion technologies that have historically been used for the destruction of POPs may fail to meet the stringent environmental standards or DRE requirements established for POPs if the incinerator is not operated under stringent technical requirements.

In addition, site owners and operators, remedial project managers, and other interested parties have expressed concern about the potential environmental and health effects associated with combustion of POPs. One concern arises because combustion technologies can create polychlorinated dibenzo-p-dioxins (dioxins) and polychlorinated dibenzo-p-furans (furans). Dioxins and furans have been characterized by US EPA as "possible" human carcinogens and are associated with serious human health problems (http://www.atsdr.cdc.gov/toxprofiles/phs104.html).

Most of the POPs-containing stockpiles are located in developing countries whose incinerators do not, in general, provide high DREs. Therefore, these developing countries must ship obsolete POPs stockpiles to developed countries for treatment and disposal. International regulations on transporting contaminated material are strict and transporting obsolete POPs from developing countries to developed countries can be cost prohibitive. Due to human health and environmental concerns associated with waste incineration, some countries (e.g., Australia and the Philippines) have non-incineration policies. Based on limitations associated with combustion technology, concerns with incineration, and an ongoing desire to find more cost effective solutions, environmental professionals are examining the application of non-combustion technologies to remediate POPs in stockpiles and soil (Ref. 71).

## 1.1    Purpose of Report

This report is intended to provide a high level summary of information for federal, state, and local regulators, site owners and operators, consultants, and other stakeholders on the applicability of existing and emerging, non-combustion technologies for the remediation of POPs in soil. The report provides short descriptions of these technologies and presents them based on the POPs treated, media treated, pretreatment requirements, performance and cost. Case studies are provided to illustrate various considerations associated with selecting a non-combustion technology. However, the report is not intended as a step-wise, or complete guide to selecting remediation technologies for POPs. This report is a second edition of a report initially published by US EPA in 2005 (EPA-542-R-05-006).

Because of international interest in POP waste management using alternate technologies, several organizations have published documents on this topic. Additional information on non-combustion technologies for the remediation of POPs waste is available in documents presented in Section 2.8 of this report. Some of the technologies discussed in these documents have progressed from the development stage to a commercial stage; other commercial technologies discussed in these reports are no longer being developed. Also, additional promising destruction technologies for POPs have been developed since the first edition of this report and other POPs treatment technology reports were prepared. The purpose of this US EPA report is to summarize and update the older reports in a reader's guide format, with links to sources of further information.

## 1.2    Methodology

In developing the 2005 report, US EPA identified non-combustion technologies for remediation of POPs in soil by reviewing technical literature, US EPA reports, and US EPA databases such as the Federal Remediation Technologies Roundtable (FRTR) (www.frtr.gov) (Ref. 18) and the Remediation and Characterization Innovative Technologies (REACHIT) system, and by contacting technology vendors and experts in the field. For this edition of the report, US EPA contacted technology vendors listed in the 2005 report for technology updates. Additional research was conducted to locate new technologies for the treatment of all 26 POPs identified by the Stockholm Convention and LRTAP. The US EPA REACHIT system could not be searched since its use was discontinued in 2008. Limited but concise data about remediation technologies is located in US EPA's Clean-Up Information (CLU-IN) (http://www.clu-in.org/vendor/vendorinfo/). In addition, a key source of information for this report was communications with John Vijgen of the IHPA. While this report has been reviewed by experts in the field, some of the information sources cited have not been peer-reviewed.

From this research, non-combustion technologies for POPs were identified. For each technology, the following information was identified: commercial availability; the processes used; advantages and limitations; POPs treated; sites where the technology was applied at full-, pilot- or bench-scale; technology performance results; cost information; and lessons learned. This report discusses technologies that have treated one or more of the 26 POPs presented in Table 1-1. Some technologies previously discussed in other sources are no longer commercially available or have not been used to treat POPs; therefore, these technologies are not included. Technologies identified in the first edition (2005) of this report that are not currently commercially available are described in Appendix E.

Based on the available information, US EPA reviewed the types of waste and contaminants treated, and summarized the results from use of the technology. Performance data were evaluated based on the concentrations of specific POPs before- and after-treatment. For many of the specific projects

> **INTERNATIONAL HCH AND PESTICIDES ASSOCIATION** PUBLISHED 15 FACT SHEETS ABOUT EMERGING NON-COMBUSTION ALTERNATIVES FOR THE ECONOMICAL DESTRUCTION OF POPs (http://www.ihpa.info/resources/library/). THESE FACT SHEETS WERE USED AS A KEY INFORMATION SOURCE DURING DEVELOPMENT OF THIS REPORT.

described in this report, gaps existed in the information available. For example, for some projects, little or no performance data were provided. US EPA did not perform independent evaluations of technology performance to support this report. However, where feasible, data gaps were addressed by contacting specific vendors, technology users, and representatives of the IHPA.

## 1.3    Report Organization

This report includes six sections and several appendices.
- **Section 1.0** is an introduction discussing the purpose, methodology, and organization of the report.
- **Section 2.0** provides background information about international treaties and organizations that address POPs issues and about the sources, characteristics, and health effects of POPs, including chemical structures and toxicology profiles.
- **Section 3.0** presents technology overviews; more detailed information for some technologies is then provided in technology-specific fact sheets in the appendices to this report. Seventeen technologies for POP treatment are described in Section 3.0, organized into three subsections based on the scale of application. Section 3.1 contains descriptions of full-scale technologies that have treated POPs. Section 3.2 and Section 3.3 contain descriptions of pilot-scale and bench-scale technologies, respectively, that have been tested on POPs.
- **Section 4.0** lists web-based information sources used to prepare this report.
- **Section 5.0** contains contact details for technology vendors.
- **Section 6.0** lists references used in the preparation of this report.
- **Appendix A** provides chemical structures, uses and effects of POPs listed under the Stockholm Convention and LRTAP.
- **Appendices B, C, and D** provide fact sheets prepared by US EPA for anaerobic bioremediation using blood meal for the treatment of toxaphene in soil, DARAMEND®, and in situ thermal desorption (ISTD), respectively, which were modeled after the fact sheets prepared by IHPA and are described below. Fact sheets for 10 other POP treatment technologies presented in this report were previously published in "Evaluation of Demonstrated and Emerging Remedial Action Technologies for the Treatment of Contaminated Land and Groundwater (Phase III)," which was issued by the IHPA in 2002. US EPA reviewed the 10 technologies as part of work for this report, as well as three additional technologies for which fact sheets were prepared by IHPA (see list in Section 2.8). Technologies identified in the first edition (2005) of this report that are not currently commercially available are provided in Appendix E. This review was implemented to evaluate whether additional, more recent information was available for these technologies. Through 2008, four additional fact sheets for other POP treatment technologies were prepared by IHPA. In addition, other technologies in this report were updated with site-specific performance data and included in their respective sections, as appropriate.

> **FURTHER INFORMATION ABOUT NON-COMBUSTION TECHNOLOGIES FOR REMEDIATION OF POPS IS PROVIDED AT** www.clu-in.org/POPs.

- **Appendix E** provides technologies identified in the first edition (2005) of this report that are not currently commercially available.

## 2.0    BACKGROUND

This section provides background information about the Stockholm Convention, Basel Convention and LRTAP. It also provides information about the sources, characteristics, and health effects of POPs. It also identifies technology categories and documents that address the treatment of POPs.

### 2.1    Stockholm Convention on POPs

The Stockholm Convention is a global treaty intended to protect human health and the environment from POPs. As of July 2010, 184 countries and one regional economic integration organization (i.e., the European Union) are Parties to the Convention. The US signed the Stockholm Convention on May 23, 2001 but as of July 2010 has not yet ratified the Convention (Ref. 61).

The Stockholm Convention has had a large impact on various countries around the world. For example, the Stockholm Convention designates the Global Environmental Facility (GEF) as the principal entity entrusted with the operations of the financial mechanism of the Convention. The GEF was originally established in 1991 and is the largest funder of projects to improve the global environment[6]. Currently, it unites 182 member governments — in partnership with international institutions, nongovernmental organizations, and the private sector — to address global environmental issues.

The Stockholm Convention's COP has established guidance for the GEF financial mechanism that emphasizes capacity building and establishes the country-specific National Implementation Plan (NIP) as the main driver for implementation activities. Specifically, the COP recommended that resources should be allocated to activities "that are in conformity with, and supportive of, the priorities identified in [Parties'] respective NIPs. This guidance has been reaffirmed and updated at subsequent COP meetings. In sum, the GEF has distributed grants to Parties to the Convention to support their development of their NIP. The NIP will:

(1) Include an initial inventory of POP stockpiles (including their location),
(2) Provide a framework for developing national laws on POPs, and
(3) Provide an action plan that details how to prioritize POPs, monitor the POPs inventory, and design a plan to eliminate POPs (short term and long-term plans).

### 2.2    Basel Convention

The Basel Convention is a global environmental agreement that focuses on the international transportation and disposal of hazardous waste. The convention, by means of a treaty, was first put into effect in May 1992. In 2004, the Basel Convention invited signatories of the Stockholm Convention to consider the development of information on best available techniques and environmental practices with respect to POPs (Refs. 69 and 70). As of June 2010, 173 parties have either signed, or signed and ratified the treaty. The US signed the treaty on March 22, 1990, but, like the Stockholm Convention, as of June 2010 has not ratified it (Ref. 12).

---

[6] More on the GEF can be found at: http://www.thegef.org/gef/

## 2.3    Convention on Long-Range Transboundary Air Pollution (LRTAP) – Protocol on POPs

The Convention on Long-Range Transboundary Air Pollution (LRTAP [7]) was signed in 1979 by 34 governments and the European Community to address issues with air pollution on a regional basis. LRTAP entered into force in 1983 and has been implemented through eight (8) Protocols that provide specific requirements for countries to reduce air emissions and pollution. In 1998, the LRTAP adopted the Protocol on POPs that focuses on a list of 16 compounds that have been singled out according to agreed risk criteria. The US is a Party to the LRTAP Convention but has not yet ratified the POPs Protocol. The compounds consist of eleven pesticides, two industrial chemicals and three by-products/contaminants. The ultimate objective is to eliminate any discharges, emissions and losses of POPs. As of April 2010, 51 parties had ratified LRTAP. The US is a signatory to the LRTAP's Protocol on POPs – but has not yet ratified the Protocol.

## 2.4    Sources of POPs

Most POPs originate from man-made sources associated with the production, use, and disposal of certain organic chemicals. Some POPs are intentionally produced, while others are unintentional by-products of industrial processes or result from the combustion of organic chemicals. The 24 POPs currently within the scope of the Stockholm Convention (or under review) include 14 pesticides and 10 industrial chemicals or by-products (Ref. 24). Table 1-1 lists these POPs.

The 14 pesticides targeted by the Stockholm Convention were produced intentionally and used on agricultural crops or for public health vector control. Over time, significant human health and environmental impacts were identified for these pesticides. By the late 1970s, these pesticides had been banned or subjected to severe use restrictions in many countries. However, some of these 14 pesticides are still used in parts of the world where they are considered essential for protecting public health (Ref. 24).

The 10 industrial chemicals and by-product POPs within the scope of the Stockholm Convention include PCBs, dioxins, furans, brominated flame retardants (BFRs), PFOS and pentachlorobenzene.

PCBs were produced intentionally but typically have been released into the environment unintentionally. The most significant use of PCBs was as a dielectric fluid (a fluid which can sustain a steady electrical field and act as an electrical insulator) in transformers and other electrical and hydraulic equipment. Most countries stopped producing PCBs in the 1980s; for example, equipment manufactured in the US after 1979 usually does not contain PCBs. However, older equipment containing PCBs is still in use. Most capacitors manufactured in the US before 1979 also contained PCBs.

Dioxins and furans are usually produced and released unintentionally. They may be generated by industrial processes or by combustion, including fuel burning in vehicles, municipal and medical waste incineration, open burning of trash, and forest fires (Ref. 24).

_____

[7] More on LRTAP can be found at:  http://www.unece.org/env/lrtap/

## 2.5    Characteristics and Health Effects of POPs

POPs are synthetic chemicals with the following properties (Ref. 24):
- They are toxic and can have adverse effects on human health and animals.
- They are chemically stable and do not readily degrade in the environment.
- They are lipophillic (possessing an affinity for fats) and easily soluble in fat.
- They accumulate and biomagnify as they move through the food chain.
- They move over long distances in nature and can be found in regions far from their points of manufacture, use, or disposal.

POPs are associated with serious human health problems, including cancer, neurological damage, birth defects, sterility, and immune system defects.  US EPA has classified certain chemicals as "probable" human carcinogens[8], including aldrin, alpha- and beta-HCH, dieldrin, chlordane, DDT, heptachlor, HCB, toxaphene, chlordecone, lindane (i.e., gamma-HCH), dioxins and furans, HCBD and PCBs.  Laboratory studies have shown that low doses of certain POPs can adversely affect organ systems.  Chronic

> **FURTHER INFORMATION ABOUT THE TOXICOLOGICAL AND CHEMICAL PROPERTIES OF POPs IS AVAILABLE AT HTTP://WWW.ATSDR.CDC.GOV/**

exposure to low doses of certain POPs may affect the immune and reproductive systems.  Exposure to high levels of certain POPs can cause serious health effects or death.  The primary potential human health effects associated with certain POPs are listed below (Refs. 16 and 24):

- Cancer
- Immune system suppression
- Nervous system disorders
- Reproductive damage
- Altered sex ratio
- Reduced fertility
- Birth defects
- Liver, thyroid, kidney, blood, and immune system damage
- Endocrine disruption
- Developmental disorders
- Shortened lactation in nursing women
- Chloracne and other skin disorders

In addition, studies have linked POP exposure to diseases and abnormalities in a number of wildlife species, including various species of fish, birds, and mammals.  For example, in certain birds of prey, high levels of DDT caused eggshells to thin to the point that the eggs could not produce live offspring (Ref. 24).

Table 2-1 provides toxicological and chemical properties of the POPs listed and under review by the Stockholm Convention and LRTAP.  Appendix A provides chemical structures, uses and effects of chemicals listed under the Stockholm Convention and LRTAP.

---

[8] Based on the 1986 USEPA classification of carcinogens, "probable" carcinogens (Group B) include those agents for which the weight of evidence of human carcinogenicity based on epidemiological studies is "limited" and those agents for which the weight of evidence of human carcinogenicity based on animal studies is "sufficient" (Ref. 56).

## Table 2-1. Toxicology and Chemical Properties of POPs Listed and Under Review by the Stockholm Convention and LRTAP

| POPs | Molecular Formula | $LD_{50}$ (mg/kg) | Half Life | Water Solubility (mg/L) | Solubility in other Solvents | Log $K_{ow}$ | Vapor Pressure@ 25°C (mm Hg) | Log $K_{oc}$ | Henry's Law Constant (dimensionless) |
|---|---|---|---|---|---|---|---|---|---|
| *Pesticides* | | | | | | | | | |
| Aldrin | $C_{12}H_8Cl_6$ | 39.0-64.0 | 53 days | 0.01 | Very soluble in most organic solvents | 6.50 | $1.20 \times 10^{-4}$ | 7.67 | $4.90 \times 10^{-5}$ |
| Alpha-Hexachlorocyclohexane (HCH) | $C_6H_6Cl_6$ | 1,000-4,000 | 54.4-56.1 days | 6.95 | Soluble in alcohol, ethanol and ether | 3.80 | $4.50 \times 10^{-5}$ | 3.57 | $6.86 \times 10^{-6}$ |
| Beta-HCH | $C_6H_6Cl_6$ | <900 | 100-184 days | 5.00 | Soluble in ethanol, ether and benzene | 3.78 | $3.60 \times 10^{-7}$ @20°C | 3.57 | $4.50 \times 10^{-7}$ |
| Chlordane | $C_{10}H_6Cl_8$ | 83.0-590 | 93.2-154 days | 0.06 | Miscible in hydrocarbon solvents | 5.54 | $2.20 \times 10^{-5}$ | 3.49-4.64 | $4.85 \times 10^{-5}$ |
| Chlordecone | $C_{10}Cl_{10}O$ | 91.3-132 | 10 days | 3.00 | Soluble in hydrocarbon solvents, alcohols and ketones | 4.50 | $3.00 \times 10^{-7}$ | 3.38-3.42 | $2.50 \times 10^{-8}$ @ 22°C |
| Dichlorodiphenyl Trichloroethane (DDT) | $C_{14}H_9Cl_5$ | 45.0-63.0 | 22 days-1 year | 0.03 | Slightly soluble in ethanol, very soluble in acetone and ethyl ether | 6.91 | $1.60 \times 10^{-7}$ @ 20°C | 5.18 | $8.30 \times 10^{-6}$ |
| Dieldrin | $C_{12}H_8Cl_6O$ | 37-46 | 5 years | 0.05 | Soluble in most organic solvents, except aliphatic petroleum solvents and methyl alcohol | 6.20 | 6.00 | 6.67 | $5.20 \times 10^{-6}$ |
| Endosulfan | $C_9H_6Cl_6O_3S$ | 40-121 | 39.5-42.1 days | 0.060-0.100 | Soluble in dichloromethane, ethanol, ethyl acetate, hexane, toluene, acetone, benzene, carbon tetrachloride, chloroform, ethanol, kerosene, methanol, xylene | 3.55-3.62 | $1.00 \times 10^{-5}$ | 3.5 | $1.00 \times 10^{-5}$ |
| Endrin | $C_{12}H_8Cl_6O$ | 7.0-43.0 | 14 years | 0.20 | Soluble in acetone, benzene, carbon tetrachloride, hexane and xylene | 5.34-5.60 | $2.00 \times 10^{-7}$ | 4.53 | $4.00 \times 10^{-7}$ |
| Heptachlor | $C_{10}H_5Cl_7$ | 39-144 | 38-44.8 days | 0.05 | Soluble in most organic solvents | 6.10 | $3.00 \times 10^{-4}$ | 4.34 | $2.94 \times 10^{-4}$ |
| Hexachlorobenzene (HCB) | $C_6Cl_6$ | 1,700-4,000 | 53 days | 0.01 | Insoluble in water, slightly soluble in ethanol, very soluble in benzene | 6.50 | $1.20 \times 10^{-4}$ | 7.67 | $4.90 \times 10^{-5}$ |

## Table 2-1. Toxicology and Chemical Properties of POPs Listed and Under Review by the Stockholm Convention and LRTAP

| POPs | Molecular Formula | LD$_{50}$ (mg/kg) | Half Life | Water Solubility (mg/L) | Solubility in other Solvents | Log K$_{ow}$ | Vapor Pressure@ 25°C (mm Hg) | Log K$_{oc}$ | Henry's Law Constant (dimensionless) |
|---|---|---|---|---|---|---|---|---|---|
| Lindane (Gamma - HCH) | $C_{12}H_4Br_6$ | 900-1,000 | 3-6 years | 0.01 | Soluble in ether and benzene | 5.73 | $1.09 \times 10^{-5}$ @ 20°C | 6.08 | $5.8 \times 10^{-4}$ |
| Mirex | $C_{10}Cl_{12}$ | 365-740 | 62.1-107 days | 17.00 | Soluble in dioxane, xylene, benzene, methyl ethyl ketone | 3.72 | $4.20 \times 10^{-5}$ @ 20°C | 3.57 | $3.50 \times 10^{-6}$ |
| Toxaphene | $C_{10}H_{10}Cl_8$ | 80-293 | 10 days | 0.60 | Freely soluble in aromatic hydrocarbons, readily soluble in organic solvents including petroleum oils | 5.28 | $3.00 \times 10^{-7}$ | 3.76 | $5.16 \times 10^{-4}$ @ 22°C |
| *Industrial Chemicals or By-Products* | | | | | | | | | |
| Polychlorinated biphenyls (PCB) | $C_{12}H_{10-x}Cl_x$ | 1,010-4,250 | 9 years | 0.42 | Very soluble in organic solvents | 5.60 | $4.00 \times 10^{-4}$ | NA | $2.90 \times 10^{-4}$ |
| Dioxins (numerical data for tetrachloro-dibenzo-p-dioxin) | $C_{12}H_xCl_xO_2$ | 0.022-0.045 | 7-12 years | 0.001 | Soluble in dichlorobenzene, chlorobenzene, benzene, chloroform and n-octanol | 7.02-8.70 | $7.50 \times 10^{-9}$ | NA | $1.61 \times 10^{-5}$ - $1.02 \times 10^{-4}$ |
| Furans | $C_4H_4O$ | 0.916 | 2.6 days | 0.010 | Soluble in toluene | 4.00-5.00 | No data | NA | NA |
| Hexabromobiphenyl | $C_{12}H_4Br_6$ | 65-149 | >6 months | 0.011 | Soluble in acetone and benzene | 6.39 | $5.20 \times 10^{-8}$ | 3.33-3.87 | $3.90 \times 10^{-6}$ |
| Octabromodiphenyl ether | $C_{12}H_2Br_8O$ | 65-149 | 76 days[2] | 0.0005 | Soluble in acetone, methanol and benzene | 6.29 | $6.59 \times 10^{-6}$ @ 21°C | NA | 10.6 |
| Pentabromodiphenyl ether | $C_{12}H_5Br_5O$ | 65-149 | 150 days | 0.013 | Soluble in methanol, miscible in toluene | 6.64-6.97 | $2.20 \times 10^{-7}$ - $5.50 \times 10^{-7}$ | 4.89-5.10 | $1.20 \times 10^{-5}$ |
| Pentachlorobenzene | $C_6HCl_5$ | 33-330 | 260-7300 days | 0.56 | Low solubility in water | 4.88-6.12 | $1.65 \times 10^{-2}$ | 6.08 | $5.8 \times 10^{-4}$ |
| Perfluorooctane sulfonate (PFOS) | $C_8F_{17}SO_3$ | 199-318 | >41 years[3] | 519-680 | Soluble in ethanol and methanol | NA | $2.40 \times 10^{-6}$ | 2.57 | $3.09 \times 10^{-9}$ |
| Hexachlorobutadiene (HCBD) | $C_4Cl_6$ | 200-580 | 1.6 years[5] | 2.00-2.55 @ 20°C | Soluble in ethanol and ether | 4.78 | 0.15 | 3.67 | 0.001-0.026 |
| Hexabromocyclo-dodecane (HBCDD) | $C_{12}H_{18}Br_6$ | 500-1,000 | 66-101 days[4] | 0.066 | Low water solubility | 5.62 | $4.70 \times 10^{-6}$ | NA | NA |

## Table 2-1. Toxicology and Chemical Properties of POPs Listed and Under Review by the Stockholm Convention and LRTAP

| POPs | Molecular Formula | LD$_{50}$ (mg/kg) | Half Life | Water Solubility (mg/L) | Solubility in other Solvents | Log K$_{ow}$ | Vapor Pressure@ 25°C (mm Hg) | Log K$_{oc}$ | Henry's Law Constant (dimensionless) |
|---|---|---|---|---|---|---|---|---|---|
| Polychlorinated Naphthalenes (PCN) | C$_{10}$H$_{10-n}$Cl$_n$ | 530-710 | 2-12 days | 31.7 | Soluble in benzene, alcohol, ether and acetone | 3.29-3.37 | 0.087 | 2.97 - 3.27 | 4.6x10$^{-4}$ |
| Short-chained chlorinated paraffins (SCCP) | C$_x$H$_{(2x-y+2)}$Cl$_y$ | 0.34 | >1 year | 0.003-0.994 | Soluble in chlorinated solvents, aromatic hydrocarbons, ketones, esters, ethers, mineral oils and some cutting oils | 4.48-8.69 | 2.10x10$^{-9}$ - 1.88x10$^{-2}$ | NA | 0.10-18.0 |

Notes:

1: Data and definitions used in this table are derived from the Agency for Toxic Substances and Disease Registry (ATSDR) website at http://www.atsdr.cdc.gov and the Stockholm Convention on Persistent Organic Pollutants website at http://chm.pops.int/.

2: Half life for Octabromodiphenyl ether in air.

3: Half life for Perfluorooctane sulfonate in water.

4: Half life for Hexabromocyclododecane in sediments

5: Half life for Hexachlorobutadiene in air.

mg/kg = milligram per kilogram

LD$_{50}$ = Lethal Dose 50% is the dose of a substance required to kill 50% of the exposed test subjects.

Half Life = The rate at which a chemical breaks down is usually defined by how long it takes for half of the chemical to break down.

Log K$_{ow}$ = The octanol/water partition coefficient is used as a measurement of a compound's bioaccumulation potential.

mm Hg = millimeters of mercury (unit for standard air pressure)

mg/L = milligram per liter

Log K$_{oc}$ = The organic carbon partition coefficient is used as a measurement of soil adsorption potential.

Henry's Law Constant = A measurement that is used to estimate the tendency of a chemical to partition between its vapor phase and water.

NA = Not available

## 2.6 Review of Chemical Characteristics of POPs Listed and Under Review for the 2009 Stockholm Convention

To determine if the POPs listed and under review in the Stockholm Convention of 2009 would be amenable to treatment using similar non-combustion technologies identified for the POPs listed by the 2001 Stockholm Convention, these new POPs were grouped and compared. For classification purposes, physical organic chemistry principles and "structure-activity relationships" (which is a major tool for new drug development) are used in this analysis. Both fields of analysis are based on types of constituents, structures, and reaction rates (Ref. 14 and 49).

### *Chemicals Added at the May 2009 Stockholm Convention Conference of Parties (COP) meeting*

### Hexachlorocyclohexane (HCH) isomers:  Lindane (i.e., Gamma-HCH), Alpha-HCH, and Beta-HCH

Based on the types of chemical constituents, structures, and reaction rates, Lindane (i.e., gamma-HCH) and 2 other HCH isomers (i.e., alpha-HCH and beta-HCH) will react with other chemicals and produce combustion products much like toxaphene (Ref. 14 and 49).  Complete combustion products are expected to include the usual organic compound products (carbon dioxide and water) and hydrochloric acid. Incomplete combustion products include carbon monoxide, acrolein, phosgene, chlorinated dioxins and chlorinated furans.  The products of non-combustion chemical technologies will depend on proprietary chemicals and their reactions under the specific treatment conditions of the technology.

### Chlordecone

Chlordecone (commonly know as its tradename Kepone®) is an isomer of mirex.  Based on the types of chemical constituents, structures, and reaction rates, chlordecone will react essentially identically to mirex (Ref. 14 and 49). Complete combustion products are expected to include the usual organic compound products (carbon dioxide and water) and hydrochloric acid.  Incomplete combustion products include carbon monoxide, acrolein, phosgene, chlorinated dioxins and chlorinated furans. The products of non-combustion chemical technologies will depend on proprietary chemicals and their reactions under the specific treatment conditions of the technology.

### Brominated compounds (octabromodiphenyl ether, penta-BDE, and hexabromobiphenyl)

Based on the types of chemical constituents, structures, and reaction rates, these three brominated compounds will react similarly to PCBs (Ref. 14 and 49).  However, these brominated compounds will probably be more reactive (less time/energy required for given amount of reaction) since bromine is a better "leaving group" than chlorine. A "leaving group" is the atom or functional group that breaks its bond with a carbon atom during the reaction.  Complete combustion products are expected to include the usual organic compound products (carbon dioxide and water) and hydrobromic acid.  Incomplete combustion products include carbon monoxide, carbonyl bromide, and brominated dioxins and furans. The products of non-combustion chemical technologies will depend on proprietary chemicals and their reactions under the specific treatment conditions of the technology.

### Pentachlorobenzene

Based on the types of chemical constituents, structures, and reaction rates, pentachlorobenzene will be very similar to HCB (Ref. 14 and 49).  Complete combustion products are expected to include the usual organic compound products (carbon dioxide and water) and hydrochloric acid.  Incomplete combustion products include carbon monoxide, acrolein, phosgene, chlorinated dioxins and chlorinated furans.  The products of non-combustion chemical technologies will depend on proprietary chemicals and their reactions under the specific treatment conditions of the technology.

## Perfluorooctane sulfonate (PFOS)

PFOS is the only new POP with no close similarity to any of the previously listed POPs. Based on the types of chemical constituents, structures, and reaction rates, it would undergo the same reactions as chlordane, lindane, toxaphene, and other aliphatic chlorinated compounds, but will be considerably less reactive (more time/energy required for a given amount of reaction), since fluorine is a very poor "leaving group" (Ref. 14 and 49). There is also one special case: PFOS is relatively water-soluble, especially in alkaline environments. Therefore a base-catalyzed reaction in aqueous media may proceed relatively rapidly because the PFOS is more available to the other reactants. In contrast, other non-combustion chemical technologies will be much less effective with PFOS than with previously discussed POPs. Complete combustion products are expected to include the usual organic compound products (carbon dioxide and water), hydrofluoric and sulfuric acids. Incomplete combustion products include carbon monoxide, carbonyl difluoride, sulfur oxides ($SO_x$) and fluorinated dioxins and furans.

### *Chemicals under Review at the May 2009 Stockholm Convention Conference of Parties (COP) meeting*

## Endosulfan

Based on the types of chemical constituents, structures, and reaction rates, endosulfan will be similar to aldrin/dieldrin (Ref. 14 and 49). It includes a sulfur atom, but that will be relatively labile; therefore, the sulfur atom should have no real effect on the properties that affect decomposition to a less toxic compound. Complete combustion products are expected to include the usual organic compound products (carbon dioxide and water), hydrochloric and sulfuric acids. Incomplete combustion products include carbon monoxide, sulfur oxides ($SO_x$), phosgene, and chlorinated dioxins and furans. The products of non-combustion chemical technologies will depend on proprietary chemicals and their reactions under the specific treatment conditions of the technology.

## Hexabromocyclododecane (HBCDD)

Based on the types of chemical constituents, structures, and reaction rates, hexabromocyclododecane (HBCDD) will be like PCB, but probably a bit easier to break down, similar to the other brominated compounds (Ref. 14 and 49). Complete combustion products are expected to include the usual organic compound products (carbon dioxide and water) and hydrochloric acid. Incomplete combustion products include carbon monoxide, carbonyl dibromide, and brominated dioxins and furans. The products of non-combustion chemical technologies will depend on proprietary chemicals and their reactions under the specific treatment conditions of the technology.

## Short-Chained Chlorinated Paraffins (SCCP)

Based on the types of chemical constituents, structures, and reaction rates, the chlorinated paraffins will be most like toxaphene, with similarity depending on factors such as the ratio of chlorine to carbon atoms and the overall size of the molecule (Ref. 14 and 49). Complete combustion products are expected to include the usual organic compound products (carbon dioxide and water) and hydrochloric acid. Incomplete combustion products include carbon monoxide, phosgene, and chlorinated dioxins and furans. The products of non-combustion chemical technologies will depend on proprietary chemicals and their reactions under the specific treatment conditions of the technology.

## 2.7    Treatment of POPs

As mentioned before, POP-contaminated soil has been widely treated using combustion systems employing high temperature incineration to destroy the contaminants. Incineration is widely used because high-temperature incinerators can address large volumes of contaminated material and can treat most contaminants. Though incineration can be used to treat POPs, there are several limitations associated with this technology, as discussed in Section 1.0. Other technology categories that can be used to treat POPs include: (1) thermal desorption and degradation, (2) chemical degradation, (3) physical-chemical degradation, (4) thermal-chemical degradation, (5) biodegradation, and (6) phytoremediation. Technologies under these categories are discussed further in Section 3.0.

## 2.8    Related Documents

Three organizations, UNEP, Africa Stockpiles Programme and IHPA, have developed summary/overview reports and fact sheets about non-combustion technologies for POPs treatment. These documents are provided below, with a list of the technologies addressed by each report.

- IHPA, 2002. IHPA and North Atlantic Treaty Organization (NATO) Committee on the Challenges of Modern Society (CCMS) Pilot Study Fellowship Report: "Evaluation of Demonstrated and Emerging Remedial Action Technologies for the Treatment of Contaminated Land and Groundwater (Phase III)." Online Address: http://www.ihpa.info/resources/library/. This report (Ref. 44) describes emerging non-combustion alternatives for the destruction of POPs. Mr. John Vijgen of IHPA gathered the technology data and prepared the report and the fact sheets for the 12 technologies listed below:

    1. Base-catalyzed decomposition (BCD)
    2. CerOx$^{TM}$
    3. Gas-phase chemical reduction process (GPCR)
    4. GeoMelt$^{TM}$
    5. In situ thermal destruction
    6. Mechanochemical dehalogenation (MCD$^{TM}$)
    7. Plasma arc (PLASCON$^{R}$)
    8. Self-propagating high-temperature dehalogenation (SPHTD)
    9. Silver II$^{TM}$
    10. Solvated electron technology
    11. Supercritical Water Oxidation (SCWO)
    12. TDT-3R$^{TM}$

- IHPA, 2009. Provisional Fact Sheets prepared by IHPA (POPs Technology Specification and Data Sheets) for the Secretariat of the Basel Convention. IHPA prepared and updated six fact sheets describing non-combustion technologies in 2009. The six technologies are listed below:

    1. Catalytic hydrodechlorination (CHD)
    2. Potassium tert-butoxide (t-BuOK) method
    3. GeoMelt$^{TM}$
    4. Supercritical water oxidation (SCWO)
    5. Radicalplanet Technology (Mechanochemical Principle)
    6. Waste to gas conversion

- UNEP, Science and Technology Advisory Panel (STAP) of the Global Environmental Facility (GEF). 2004. "Review of Emerging, Innovative Technologies for the Destruction and Decontamination of POPs and the Identification of Promising Technologies for Use in Developing Countries." GF/8000-02-02-2205. January. Online Address: http://www.basel.int/techmatters/review_pop_feb04.pdf. This report (Ref. 72) provides a summary overview of non-combustion technologies that are considered to be innovative and emerging and that have been identified as potentially promising for the destruction of POPs in soil. The report was a background document for the STAP-GEF workshop held in Washington, DC, in October 2003 and was based on work by the International Centre for Sustainability Engineering and Science, Faculty of Engineering, at the University of Auckland, New Zealand. The report contains overviews of the following 27 non-combustion technologies:

1. BCD
2. Bioremediation/Fenton reaction
3. Catalytic hydrogenation
4. DARAMEND® bioremediation
5. Enzyme degradation
6. Fe (III) photocatalyst degradation
7. GPCR
8. GeoMelt™ process
9. In situ bioremediation of soils
10. MCD
11. Mediated electrochemical oxidation (AEA Silver II)
12. Mediated electrochemical oxidation (CerOx™)
13. $MnO_x/TiO_2 - Al_2O_3$ catalyst degradation
14. Molten metal
15. Molten salt oxidation
16. Molten slag process
17. Ozonation/electrical discharge destruction
18. Photochemically enhanced microbial degradation
19. Phytoremediation
20. Plasma arc (PLASCON™)
21. Pyrolysis
22. SPHTD
23. Sodium reduction (SR)
24. Solvated electron technology
25. SCWO
26. $TiO_2$ – based $V_2O_5/WO_3$ catalysis
27. White rot fungi bioremediation

- The International Centre for Science and High Technology - United Nations Industrial Development Organization, 2007. "Non-Combustion Technologies for POPs Destruction – Review and Evaluation." Trieste, Italy. March. Online Address: www.ics.trieste.it. This report (Ref. 42) provides information about alternative non-combustion technologies for the treatment of POPs. The report contains summaries for the following 15 technologies:

1. Ball Milling – MCD and DMCR
2. BCD
3. Cerox™
4. Geomelt™
5. GPCR™
6. HydroDec™
7. MSO
8. PACT™
9. PLASCON^R
10. PWC™
11. SCWO
12. SET™
13. Silver II™
14. SPHTD
15. SR

- Africa Stockpiles Programme, 2008. "Review and Update of Technology." Online Address: http://www.africastockpiles.net/ This report provides an overview of various non-combustion technologies and includes fact sheets for the seven technologies listed below:

1. BCD
2. GPCR
3. Plasma arc (PLASCON^R)
4. SCWO
5. GeoMelt™
6. Ball Milling (Radical Planet)
7. Thermopower (Thermal Retorting) Process

- Japan Environmental Safety Corporation (JESCO), 2005. JESCO is a primary technology provider for the treatment of PCB contaminated wastes. Dr. Noma of National Institute for

Environmental Studies developed fact sheets for the following six technologies; the fact sheets are available at http://www.ihpa.info/resources/library/

1. Radicalplanet® (Mechanochemical Principle)
2. SP process (Sodium Powder Dispersion Dechlorination Process)
3. Sub-critical water oxidation
4. Supercritical Water Oxidation of Organo Corporation
5. Supercritical Water Oxidation of Kurita Industries
6. Vacuum Heating Decomposition

- Basel Convention, 2005. "Destruction and Decontamination Technologies for PCBs and Other POPs Wastes (Part III Annexes) A Training Manual for Hazardous Waste Project Managers, Volume C." Online address: http://www.basel.int/meetings/sbc/workdoc/TM-A.pdf . This report contains seven fact sheets prepared by IHPA (listed as POP Technology Specification and Data Sheets) for the Secretariat of Basel Convention. Four of these published fact sheets, listed below, pertain to non-combustion technologies for the treatment of POPs:

1. Alkali metal reduction
2. Base-catalyzed decomposition (BCD)
3. Gas-phase chemical reduction (GPCR)
4. Plasma Arc (PLASCON)

- Basel Convention, 2007. "Updated general technical guidelines for the environmentally sound management of wastes consisting of, containing or contaminated with persistent organic pollutants (POPs)" Online address: http://www.basel.int/pub/techguid/tg-POPs.pdf.  This report contains summaries for the following technologies for the treatment of POPs.

1. Alkali metal reduction
2. Base-catalysed decomposition (BCD)
3. Catalytic hydrodechlorination (CHD)
4. Photochemical dechlorination (PCD) and catalytic dechlorination (CD) reaction
5. Gas-phase chemical reduction (GPCR)
6. Plasma arc
7. Potassium tert-butoxide (t-BuOK) method
8. Supercritical water oxidation (SCWO) and subcritical water oxidation
9. Waste to gas conversion

## 3.0 NON-COMBUSTION TECHNOLOGIES

This section provides a review of selected non-combustion technologies for POPs remediation, including their implementation at both domestic and international sites. In this report, POPs include the 26 chemicals listed or under review in the Stockholm Convention on POPs and/or the LRTAP's Protocol on POPs. Non-combustion technologies are defined as processes that operate in a starved or ambient oxygen atmosphere (including thermal processes). For this report, treatment technology is defined as the primary process through which contaminant destruction occurs. Pretreatment is defined as any process that precedes the primary treatment technology to prepare the contaminated material for treatment, typically via transfer of contaminants from one media/phase to another (e.g., solid to liquid phase).

Tables 3-1 and 3-2 list the technologies addressed in this report and summarize available technology-specific information, including: capability to handle waste strength, whether treatment is ex situ or in situ, scale, contaminant treated, cost, pre-treatment needs, power requirements, technology configuration, and location of any fact sheets available for the technology. Waste strength refers to high- and low-strength wastes. High-strength waste includes soil contaminated with high concentrations of POPs. Low-strength waste includes soil contaminated with low concentrations of POPs. Table 3-1 provides information about full-scale[9] technologies and Table 3-2 provides information about pilot-scale[10] and bench-scale[11] technologies for treatment of POPs. Table 3-3 presents performance data for the technologies. The performance data include site location, contaminants treated, untreated and treated contaminant concentrations, and percent reduction of the contaminants (as available). Section 5.0 contains contact information for vendors of these various technologies.

### 3.1 Full-Scale Technologies for Treatment of POPs

This section presents 12 technologies that have been implemented to treat POPs at full scale. Each subsection focuses on a single technology and includes a description of the technology and information about its application at specific sites. Fact sheets developed by US EPA and IHPA provide additional details for some of these technologies and their applications. Appendix B, C, and D of this report provide fact sheets prepared by US EPA for anaerobic bioremediation using blood meal for the treatment of toxaphene in soil, DARAMEND®, and in situ thermal desorption (ISTD), respectively. Links to the IHPA fact sheets are included in the appropriate subsections of this report.

---

[9] A full-scale project involves use of a commercially available technology to treat industrial waste and to remediate an entire area of contamination.

[10] A pilot-scale project is usually conducted in the field to test the effectiveness of a technology and to obtain information for scaling up a treatment system to full scale.

[11] A bench-scale project is conducted on a small scale, usually in the laboratory, to evaluate a technology's ability to treat soil, waste, or water. Such a project often occurs during the early phases of technology development.

*Reference Guide to Non-combustion Technologies for Remediation of Persistent Organic Pollutants in Soil, Second Edition - 2010*

## Table 3-1. Summary of Full-Scale Non-combustion Technologies for Remediation of Persistent Organic Pollutants [1]

| Technology | Commercial Availability | Waste Strength [2] | Ex/In situ [3] | Contaminant(s) Treated POPs Pesticide(s) [4] | POPs PCBs | POPs Dioxin/Furans | Non-POPs [5] | Cost | Pre-Treatment | Power Requirement | Configuration | Fact Sheet |
|---|---|---|---|---|---|---|---|---|---|---|---|---|
| **Full-Scale Technologies** | | | | | | | | | | | | |
| Anaerobic bioremediation using blood meal for the treatment of toxaphene in soil and sediment | Yes | Low | Ex situ | Toxaphene, DDT | None | None | None | $130 to $271 per cubic yard (in 2007) | None | None | Transportable | Appendix B |
| Base Catalyzed Decomposition (BCD) | Yes | Low/High | Ex situ | Chlordane, Heptachlor, DDT, HCB, Lindane, HCH | Yes | Yes | PCP, herbicides, pesticides, insecticides | 1,400-1,700 Euros/ton (in 2004) $500,000 to $2 million for one reactor (physical plant facility only) | Thermal desorption Debris removal pH or moisture content adjustment | Low-High | Transportable and fixed | http://www.ihpa.info/resources/library/ |
| DARAMEND® | Yes | Low | Ex/In situ | Toxaphene, DDT, HCB, Dieldrin, a-HCH, B-HCH, Lindane | None | None | DDD, DDE, RDX, HMX, DNT, TNT, 2,4-D; 2,4,5-T Metoachlor, Atrizine | $55 per ton (ex situ), $12.50 per cubic yard (in situ), $30,000 per acre (in situ full-scale) (in 2005) | None | None | Transportable | Appendix C |
| Gas Phase Chemical Reduction (GPCR™) | No [6] | High | Ex situ | DDT, HCB, Dieldrin, Lindane, Aldrin | Yes | Yes | PAH, chlorobenzene | Capital cost estimate for two-Thermal Reduction Batch Processor plants (solid feed): $10.8M for full-scale, $5M for semi-mobile, and one estimate for one TRBP plant (liquid/gaseous feed): $10.3M for full-scale, $4.75M for semi-mobile. Minimum set-up costs: $10.5M for full-scale, $5M for semi-mobile | Thermal desorption | Low-High | Fixed and transportable | http://www.ihpa.info/resources/library/ |

*Reference Guide to Non-combustion Technologies for Remediation of Persistent Organic Pollutants in Soil, Second Edition - 2010*

## Table 3-1. Summary of Full-Scale Non-combustion Technologies for Remediation of Persistent Organic Pollutants [1]

| Technology | Commercial Availability | Waste Strength [2] | Ex/In situ [3] | Contaminant(s) Treated — POPs — Pesticide(s) [4] | PCBs | Dioxin/Furans | Non-POPs [5] | Cost | Pre-Treatment | Power Requirement | Configuration | Fact Sheet |
|---|---|---|---|---|---|---|---|---|---|---|---|---|
| Gene Expression Factor® (bioremediation) | Yes | Low | | | | | DDE | Initial cost for bench-scale study was $30,000. $30 to $60 per ton of contaminated soil depending on site conditions | None | None | Fixed and transportable | None |
| GeoMelt™ | Yes | Low/High | In/Ex situ | DDT, Chlordane, Dieldrin, Heptachlor, HCB | Yes | Yes | Metals and radioactive waste | NA | Dewatering /drying may be required | High | Fixed and transportable | http://www.ihpa.info/resources/library/ |
| Mechanochemical Dehalogenation (MCD™) | Yes[7] | Low/High | Ex situ | Aldrin, Dieldrin DDT, Lindane | Yes | No | DDD, DDE, HCH, PCP, PAHs, organic pesticides, hydrocarbons | NA | Grinding, drying | High | NA | http://www.ihpa.info/docs/library/libraryNATO.php |
| Plasma Arc (PLASCON™) | Yes[8] | Low/High | Ex situ | DDT, Chlordane, Endosulfan, Aldrin, HCB, Dieldrin, Lindane | Yes | Yes | NA | $1M for standard 150 kW plant | Thermal desorption | Low/High | Fixed and transportable | http://www.ihpa.info/resources/library/ |
| Radicalplanet® Technology | Yes[9] | Low/High | Ex situ | Chlordane, DDT, Endrin, HCH, Lindane | Yes | Yes | PCP, PCNB, PVC (Asbestos) | 2.8 million Euros for E-200 (one machine with 105 tons/y) 3.3 million Euros for E-500 (one machine with 210 tons/y) | None | Low | Fixed and Transportable | http://www.ihpa.info/resources/library/ |
| Solvated Electron Technology™ | Yes | Low/High | Ex situ | NA | Yes | Yes | Explosives, CFC, Halons | NA | Shredding/ grinding, dewatering/ drying | Moderate | Fixed and transportable | http://www.ihpa.info/docs/library/libraryNATO.php |
| Sonic Technology | Yes | Low/High | Ex situ | DDT | Yes | Yes | PAH, VOCs, Pesticides | NA | Mixing with solvent to produce a slurry | 75 kW | Transportable | None |

*Reference Guide to Non-combustion Technologies for Remediation of Persistent Organic Pollutants in Soil, Second Edition - 2010*

## Table 3-1. Summary of Full-Scale Non-combustion Technologies for Remediation of Persistent Organic Pollutants [1]

| Technology | Commercial Availability | Waste Strength [2] | Ex/In situ [3] | Contaminant(s) Treated | | | | Cost | Pre-Treatment | Power Requirement | Configuration | Fact Sheet |
| | | | | POPs | | | Non-POPs [5] | | | | | |
| | | | | Pesticide(s) [4] | PCBs | Dioxin/Furans | | | | | | |
| Thermal In Situ Thermal Desorption (ISTD) | Yes | Low/High | In situ | NA [10] | Yes | Yes | VOCs, SVOCs, oils, creosote, coal tar, gasoline, MTBE, volatile metals | $200 to $600 per cubic yard (data from 1996 to 2005) | Dewatering may be required | High | Transportable | Appendix D |

**Notes:**

1: Data in these tables are derived from various documents, vendor information, and other sources - both peer reviewed and not, provided in the later technology-specific sections.

2: Waste strength refers to high- and low-strength wastes. High-strength waste includes stockpiles of POP-contaminated materials and highly contaminated soil. Low-strength waste includes soil contaminated with low concentrations of POPs.

3: Ex/In situ refers to ex situ or in situ application of the technology.

4: Pesticides include the 13 pesticides addressed within the scope of the Stockholm Convention and LRTAP.

5: Non-POPs include contaminants outside the scope of Stockholm Convention and LRTAP.

6: Technology is not commercially available and is currently being modifying to improve its cost effectiveness

7: Technology is commercially available from EDL in Auckland, New Zealand and Tribochem in Wunstrof, Germany. No technology vendor is available in the US

8: SRL Plasma Pty. Ltd., an Australian company, is the patent holder of this technology. Technology commercially used in Japan.

9: Technology is commercially available only in Japan

10: According to TerraTherm, laboratory-scale work indicates that this technology can also effectively treat other POPs, including aldrin, dieldrin, endrin, chlordane, heptachlor, DDT, mirex, HCB, and toxaphene, but these contaminants have not yet been treated using ISTD at full or pilot scale.

HCB:    Hexachlorobenzene
HCH:    Hexachlorocyclohexane
BFRs:   Bromated Flame Retardants – Octabromobiphenyl ether. Pentabromodiphenyl ether. Hexabromobiphenyl & Hexabromocyclododecane (HBCDD)
DDD:    Dichlorodiphenyldichloroethane
DDE:    Dichlorodiphenyldichloroethylene
DDT:    Dichlorodiphenyltrichloroethane
DNT:    Di-nitro toluene
HMX:    High melting explosive, octahydro-1.3.5.7-tetranitro-1.3.5.7 tetrazocine
PCB:    Polychlorinated biphenyls
PCP:    Pentachlorophenol
PVC:    Polyvinyl chloride
CFC:    Chlorofluorocarbon
TNT:    2,4,6-Trinitrotoluene

MTBE:      Methyl tert-butyl ether
NA:        Not available
PAH:       Polycyclic aromatic hydrocarbons
Penta–CB:  Pentachlorobenzene
PCNB:      Pentachloronitrobenzene
PCNs:      Polychlorinated napthalenes
SCCPs:     Short-chained chlorinated paraffins
SVOC:      Semivolatile organic compound
VOC:       Volatile organic compound

21

*Reference Guide to Non-combustion Technologies for Remediation of Persistent Organic Pollutants in Soil, Second Edition - 2010*

**Table 3-2. Summary of Pilot/Bench-Scale Non-combustion Technologies for Remediation of Persistent Organic Pollutants [1]**

| Technology | Waste Strength [2] | Ex/In situ [3] | Contaminant(s) Treated | | | | Cost | Pre-Treatment | Power Requirement | Configuration | Fact Sheet |
| | | | POPs | | | Non-POPs [5] | | | | | |
| | | | Pesticide(s) [4] | PCBs | Dioxin/ Furans | | | | | | |
| **Pilot-Scale Technologies** | | | | | | | | | | | |
| Phytoremediation | Low | In/Ex situ | DDT, Chlordane | Yes | No | DDE | NA | None | None | Transportable | None |
| Reductive Heating and Sodium Dispersion | Low/High | Ex situ | DDT, Chlordane, Aldrin, B-HCH | Yes | Yes | PCNB | NA | None | NA | Transportable | None |
| Sub-critical Water Oxidation | NA | Ex situ | Aldrin, Dieldrin, Chlordane | Yes | Yes | BHC | NA | Extraction into a solvent | NA | Fixed and transportable | http://www.ih pa.info/resour ces/library/ |
| **Bench-Scale Technologies** | | | | | | | | | | | |
| Self Propagating High Temperature Dehalogenation | High | Ex situ | HCB | No | No | None | NA | None | NA | NA | http://www.ih pa.info/docs/li brary/libraryN ATO.php |
| TDR-3R™ | High | Ex situ | HCB | No | No | PAH, VOCs, SVOCs | NA | Thermal desorption | High | NA | http://www.ih pa.info/docs/li brary/libraryN ATO.php |

**Notes:**

1: Data in these tables are derived from various documents, vendor information, and other sources - both peer reviewed and not, provided in the later technology-specific sections.

2: Waste strength refers to high- and low-strength wastes. High-strength waste includes stockpiles of POP-contaminated materials and highly contaminated soil. Low-strength waste includes soil contaminated with low concentrations of POPs.

3: Ex/In situ refers to ex situ or in situ application of the technology.

4: Pesticides include the 13 pesticides addressed within the scope of the Stockholm Convention and LRTAP.

5: Non-POPs include contaminants outside the scope of Stockholm Convention and LRTAP.

B-HCH: beta- hexachlorocyclohexane
BHC: Hexa-Chloro Benzene (BHC/Lindane)
DDE: Dichlorodiphenyldichloroethylene
DDT: Dichlorodiphenyltrichloroethane

HCB: Hexachlorobenzene
NA: Not available
PAH: Polycyclic aromatic hydrocarbons
PCNB: Pentachloronitrobenzene

SVOC: Semivolatile organic compound
VOC: Volatile organic compound

22

## Table 3-3. Performance of Non-combustion Technologies for Remediation of Persistent Organic Pollutants [1]

| Technology | Examples of Treatment Performance [2] | | | |
| --- | --- | --- | --- | --- |
| | Site Name or Location | Contaminant | Untreated Concentration (mg/kg) [3] | Treated Concentration (mg/kg) | Percent Reduction |
| **Full-Scale Applications** | | | | | |
| Anaerobic bioremediation using blood meal for the treatment of toxaphene in soil and sediment | Gila River Indian Community, Arizona | Toxaphene | 29-34 | 4-5 | 86-93 % |
| | Gila River Boundary, Arizona | Toxaphene | 23-110 | 5-20 | 66-82% |
| DARAMEND® | T.H. Agricultural and Nutrition Superfund Site, Montgomery, Alabama | Toxaphene | 189 (Mean) | 10 (Mean) | 89% |
| | | DDT | 81 (Mean) | 9 (Mean) | 90% |
| | Former North American Transformer South Yard Area, Milpitas, California | PCBs | 156 (Max) | <1 | 99.38% |
| Gene Expression Factor | Borello Property, Morgan Hill, California | Dieldrin | 0.48 | 0.00354 (Mean) | 99.12% |
| | | Toxaphene | 6.2 | 0.130 | 97.9% |
| | Mantegani Property, South San Francisco, California | DDT | 9.0 (Max) | <0.5 | 94.5% |
| | | Dieldrin | 1.0 (Max) | 0.5 | 50% |
| | Parsons Chemical Works, Inc. Superfund Site, Grand Ledge, Michigan | DDT | 340 (Max) | <0.016 | 99% |
| | | Chlordane | 89 (Max) | <0.08 | 99% |
| | | Dieldrin | 87 (Max) | <0.016 | 99% |
| GeoMelt™ | TSCA Spokane, Spokane, Washington | PCBs | 17,860 | ND | NA |
| | | Dioxins | 38 (Max) | ND | NA |
| | Wasatch Chemical, Salt Lake City, Utah | DDT | 1,091 (Max) | ND | NA |
| | | Chlordane | 535 (Max) | ND | NA |
| | WCS-Commercial TSCA cleanup, Andrews, Texas | PCBs | 496 | ND | NA |
| Mechanochemical Dehalogenation (MCD™) | Fruitgrowers Chemical Company Site, Mapua, New Zealand | Aldrin +Dieldrin+ Lindane (ADL) | 73.245 (Mean) | 20.612 (Mean) | 71.86% |
| | | DDX (total DDT, DDD, and DDE) | 717 (Mean) | 64.8 (Mean) | 90.96% |
| Radicalplanet® Technology | Ibraki, Japan | PCBs | 42,800 | 0.01 | 99.99% |
| | Ibraki, Japan (with Geo-Environmental Protection Center) | PCBs | 75,000 ng-TEQ | 0.13 ng-TEQ | 99.99999% |
| Solvated Electron Technology | Pennsylvania Air National Guard Site, Harrisburg International Airport, Harrisburg, Pennsylvania | PCBs | 17-560 | <1 | 99.99% |
| Sonic Technology | Juker Holdings Site, Vancouver, British Columbia | PCBs (from soil) | 400-1,600 | <25 | 98.43% |
| | | PCB (concentrate in kerosene) | 46,000 (Max) | <3 | 99.99% |

## Table 3-3. Performance of Non-combustion Technologies for Remediation of Persistent Organic Pollutants [1]

| Technology | Site Name or Location | Examples of Treatment Performance[2] | | | |
|---|---|---|---|---|---|
| | | Contaminant | Untreated Concentration (mg/kg)[3] | Treated Concentration (mg/kg) | Percent Reduction |
| In Situ Thermal Desorption (ISTD) | Tanapag Village, Saipan, Northern Mariana Islands | PCBs | 10,000 (Max) | <10 | 99.9% |
| | Former South Glens Falls Dragstrip, Moreau, New York | PCBs | 5,000 (Max) | 0.8 | 99.9% |
| | Centerville Beach, Ferndale, California | PCBs | .15–860 | <0.17 | 99.98% |
| | | Dioxin/Furans | 0.0032 (Max) | 0.00006 | 99.81% |
| | Alhambra "Wood Treater", Alhambra, California | Dioxins | .0194 | <.001 | 94.85% |
| **Pilot-Scale Applications** | | | | | |
| Base Catalyzed Decomposition (BCD) | Warren County Landfill, Warren County, North Carolina | PCBs | 81,100 | <5 | 99.99% |
| | FCX Superfund Site, Statesville, North Carolina | Heptachlor | 0.648 | ND[2] | NA |
| | | Chlordane | .02 | ND[2] | NA |
| DARAMEND® | Former Agricultural Site, Florida | Dieldrin | .046 (Mean) | .015 | 67.39% |
| | Hot-Spot Treatment, Former Manufacturing Facility, Southeastern US | Toxaphene | 127.7 (Mean) | 8.7 (Mean) | 93.19% |
| In Situ Thermal Desorption (ISTD) | Missouri Electric Works, Cape Girardeau, Missouri | PCBs | 20,000 (Max) | <0.033 | 99.99% |
| | Former Mare Island Naval Shipyard, Vallejo, California | PCBs | 2,200 (Max) | <0.033 | 99.99% |
| **Bench-Scale Applications** | | | | | |
| DARAMEND® | Former obsolete pesticide warehouse, Moldova | Lindane | 17 (Mean) | 10 (Mean) | 41.18% |
| TDT-3R™ | Gare Site, Hungary | HCB | 1,215 | 0.1 | 99.99% |

**Notes:**

1: Data in this table are derived from various document, vendor information, and other sources, cited in the later technology-specific sections.

2: Treatment examples were selected to illustrate the types of treatment performance data available.

3: The concentrations are maximum concentrations unless otherwise indicated in parenthesis.

4: The specific limits for the MDL and ND were not provided in the source document.

5: Full-scale data are present in this table but available pilot-scale data can be found in Section 3 under each specific technology.

DDD: Dichlorodiphenyldichloroethane  DDE: Dichlorodiphenyldichloroethylene  DDT: Dichlorodiphenyltrichloroethane
HCB: Hexachlorobenzene  PCBs: Polychlorinated biphenyls  ND: Not detected (concentration below method detection limit)
Max: Maximum Concentration  Mean: Mean Concentration  MDL: Method detection limit
mg/kg: Milligram per kilogram  ng-TEQ/g = Nanogram Toxic Equivalent of Dioxins per gram
NA: Not available

### 3.1.1   Anaerobic Bioremediation Using Blood Meal

This technology claims to use biostimulation with amendments to promote degradation of toxaphene in soil or sediment by native anaerobic microorganisms. For treatment, biological amendments such as blood meal (dried and powdered animal blood), which is used as a nutrient, and phosphates, which are used as a pH buffer are added to the contaminated material (Ref. 3). In some applications, starch is also used. The contaminated soil is mixed with the amendments and water. The technology can use several methods to produce soil-amendment mixtures, including blending in a dump truck, mechanical mixing in a pit, and mixing in a pug mill. The soil mixture is transferred to a lined cell, and water is added to produce a slurry. Up to a foot of water cover is provided above the settled solids. The water cover is intended to minimize the transfer of atmospheric oxygen to the soil amendment mixture so that anaerobic conditions are maintained. The lined cell is covered with a plastic sheet and the slurry is incubated for several months.

The slurry is then sampled periodically to measure contaminant concentration. The process continues until the treatment goals are achieved, at which time the cell is drained. The treated slurry is usually left in the cell;

> **THE FACT SHEET PREPARED BY US EPA IS INCLUDED IN APPENDIX B.**

however, the slurry may be dried and used as fill material on site or as a source of microorganisms for other applications of the technology.

> **TECHNOLOGY TYPE: BIODEGRADATION**
>
> **POPs TREATED: TOXAPHENE**
>
> **MEDIUM: SOIL AND SEDIMENT**
>
> **PRETREATMENT: NONE**
>
> **RESIDUALS: LOW CONCENTRATIONS OF TOXAPHENE AND CAMPHENES WITH VARYING DEGREES OF CHLORINATION**
>
> **COSTS: $130 TO $271 PER CUBIC YARD (COST IN 2007 USD)**
>
> *FULL SCALE*
> *EX SITU*

Anaerobic bioremediation using blood meal has been used to treat low-strength waste contaminated with toxaphene. Essential components such as mixing troughs are typically constructed and left in place. Other components such as mixing equipment and biological amendments have been procured locally.

The technology has been used to treat toxaphene at several livestock dip vat sites and one site with a pesticide spill. Dip vats are trenches with a pesticide formulation used to treat livestock infested with ticks. In 2007, cleanup costs in US Dollar (USD) for full-scale implementation ranged from $130 to $271 per cubic yard (Ref. 29). Performance data from applications at nine dip vat sites and one pesticide site are presented in Table 3-4.

Anaerobic bioremediation using blood meal was developed by US EPA's Environmental Response Team (ERT). The technology has been used at sites with toxaphene contaminated soil and sediments. Bench scale testing is recommended to determine if the technology will be effective at a particular site, as well as to evaluate amendment types and quantities, possible removal, and whether degradation products are formed and persist. Differences between unamended live, killed, and amended live units may also help to evaluate treatment effectiveness. Bench testing should be conducted in gas tight units to reduce volatile losses. Based on structural similarity of toxaphene to other POPs described in section 2.6 of this report, this technology may potentially be used to treat other POPs. However, because of the specificity of biochemical reactions, this technology may or may not be effective in treating other similar POPs. This technology is publicly available and is currently not patented (Ref. 4). The most recent application was in 2004 at the Gila River Boundary (GRB) site in Laveen, Arizona. Further technology information can be obtained by contacting the technology developer using the information provided in Section 5.0.

**Table 3-4. Performance of Anaerobic Bioremediation Using Blood Meal for Toxaphene Treatment**

| Site | Location | Period (Days) | Quantity of Soil Treated | | Untreated Concentration (mg/kg) | Treated Concentration (mg/kg) |
|---|---|---|---|---|---|---|
| **Gila River Indian Community (GRIC)** | | | | | | |
| GRIC Cell 1 | | 272 | | Full | 59 | 4 |
| GRIC Cell 2 | Chandler, Arizona | 272 | 3,500 cy | Full | 31 | 4 |
| GRIC Cell 3 | | 272 | | Full | 29 | 2 |
| GRIC Cell 4 | | 272 | | Full | 211 | 3 |
| **Navajo Vats Chapter** | | | | | | |
| Laahty Family Dip Vat | Zuni Nation, New Mexico | 31 | 253 cy | Full | 29 (Mean) | 4 |
| Henry O Dip Vat | Zuni Nation, New Mexico | 68 | 660 cy | Full | 23 (Mean) | 8 |
| Nazlini | NA | 108 | 3.5 tons | Pilot | 291 | 71 |
| Whippoorwill | NA | 110 | 3.5 tons | Pilot | 40 | 17 |
| Blue Canyon Road | NA | 106 | NA | NA | 100 | 17 |
| Jeddito Island | NA | 76 | NA | NA | 22 | 3 |
| Ojo Caliente | Zuni Nation, New Mexico | 14 | 200 cy | NA | 14 | 4 |
| Poverty Tank | NA | 345 | NA | NA | 33 | 8 |
| **Gila River Boundary (GRB)** | | | | | | |
| GRB (6 cells) | Laveen, Arizona | 180 | 8,000 cy | Full | 23-110 | 5-20 |

Sources:  Refs. 3 and 29

Notes:
cy = Cubic yard                                    NA =    Not available
mg/kg = Milligram per kilogram

### 3.1.2    Base-Catalyzed Decomposition

Base-Catalyzed Decomposition (BCD) is an ex situ technology that has been used to treat high-strength soil containing POP contamination.  The technology is available in both transportable and fixed configurations.

The use of BCD technology may require pre-treatment using thermal desorption when pollutants are in ppm rather than percent concentrations in contaminated matrices. Depending on the concentration of the

> **THE FACT SHEET PREPARED BY IHPA IS AVAILABLE AT** http://www.ihpa.info/resources/library/.

contaminants, a selected amount of an alkali such as sodium bicarbonate is mixed with the contaminated soil in the pre-treatment stage of the process and the mixture is heated in a thermal desorption reactor to temperatures ranging from 315 to 500°C.  The heat separates the halogenated compounds from the soil by evaporation.  In the second stage of the pre-treatment process, the volatilized contaminants pass through a condenser.  The condensate is then sent to a BCD liquid tank reactor (LTR).  Sodium hydroxide, a proprietary catalyst, and carrier oil are added to the LTR, which is then heated to above 326°C for three to six hours.  The carrier oil serves both as a suspension medium and a hydrogen donor.  The heated oil is then cooled and sampled to determine whether it meets disposal criteria.  If the oil does not meet the disposal criteria, it is returned to the LTR, reagents are added, and the reactor is reheated (Ref. 48).  The

---

| |
|---|
| **TECHNOLOGY TYPE: CHEMICAL DEGRADATION** |
| |
| **POPS TREATED: PCBS, CHLORDANE, HEPTACHLOR, DDT, HCB, HCH, LINDANE, DIOXINS, AND FURANS** |
| |
| **PRETREATMENT: THERMAL DESORPTION** |
| |
| **MEDIUM: SOIL AND LIQUIDS** |
| |
| *FULL SCALE* |
| *EX SITU* |

treated soil can be used as backfill on site. BCD was developed by US EPA's National Risk Management Research Laboratory in Cincinnati, Ohio. US EPA holds the patent rights to this technology in the US. The foreign rights are held by BCD Group Inc., Cincinnati, Ohio. The technology has been licensed by BCD Group Inc., to environmental firms in Spain, Australia, Japan, Czech Republic, and Mexico. Since its initial development in 1990, considerable technology advancements have been made with the development of a new catalyst, which reduces the reaction time in the BCD reactor (Ref. 56). This second generation technology has been applied in Australia, Mexico and Spain to treat PCB-contaminated oil.

Several full-scale applications of BCD have addressed POP-contaminated wastes. Two commercial facilities operated in Australia and treated approximately 8,000 to 10,000 tons of PCB contaminated waste, PCB-contaminated oil, 25 tons of pesticide chemicals and pesticide wastes, and 15 tons of pesticide concentrates generated and collected as a result of soil remediation. Another commercial facility has been operating in Mexico since 1998 and has treated 1,400 tons of liquids and solids contaminated with PCBs. In the Czech Republic, a full-scale BCD unit has been operating since 2006 and has treated 29,000 to 38,000 tons of soil and building debris contaminated with dioxins, furans, HCB, lindane, and HCH, as well as nearly 200 tons of waste chemicals. In addition, 300 tons of concentrated contaminants from the thermal desorption process have been treated using BCD. A system also operated between 2000 and 2002 in Spain to treat 3,500 tons of pure HCH waste. The performance data for these applications could not be obtained from the technology vendor.

BCD is a non-combustion technology that uses sodium hydroxide, a proprietary catalyst, carrier oil and heat to treat POPs contaminated soil and liquids. BCD has been used to treat PCBs, HCB, HCH, lindane, dioxins and furans. Based on structural similarity of known POPs treated using BCD to other similar POPs described in section 2.6 of this report, this technology can potentially be used to treat other POPs. However, the potential of BCD technology to treat other POPs is dependent on the proprietary chemical catalyst used and the specific reactions under the treatment conditions of this technology. BCD is licensed by BCD Group Inc. and has been used at full-scale in various countries around the world including Spain, Australia, Japan, Czech Republic and Mexico. The most recent application of this technology was in 2006 at a site in the Czech Republic. The performance data for this technology were provided by John Vijgen (IHPA). No performance data could be obtained directly from the technology vendor. Currently, no full-scale applications of this technology exist in the US. Contact information for the technology vendor is provided in Section 5.0.

### 3.1.3  DARAMEND®

DARAMEND® has been used to treat low-strength wastes contaminated with toxaphene and DDT. It is an amendment-enhanced bioremediation technology that involves the creation of sequential anoxic and oxic conditions (Ref. 54). The treatment process involves the following steps:
1. Addition of a solid-phase DARAMEND® organic soil amendment of a specific particle size distribution and nutrient profile, zero valent iron, and water to contaminated soil to produce anoxic conditions
2. Periodic tilling of the soil to promote oxic conditions
3. Repetition of the anoxic-oxic cycle until cleanup goals are achieved

The addition of the DARAMEND® organic amendment, zero valent iron, and water stimulates the biological and chemical depletion of oxygen, generating strong reducing (anoxic) conditions in the soil matrix. Diffusion of replacement oxygen into the soil matrix is prevented by near saturation of the soil pores with water. The depletion of oxygen creates a very low reduction-oxidation (redox) potential, which promotes dechlorination of organochlorine compounds. The soil matrix (contaminated soil and the amendments) is left undisturbed for the duration of the anoxic phase of the treatment cycle (typically 1 to 2 weeks). In the next (oxic) phase, periodic tilling of the soil increases diffusion of oxygen and distribution of irrigation water in the

**Bioremediation using DARAMEND® process. Ref. 1**

soil. The dechlorination products formed during the anoxic degradation process are then removed through aerobic (oxic) biodegradation processes, which are initiated and promoted by the air drying and tilling of the soil. Addition of the DARAMEND® amendment and the anoxic-oxic cycle continue until cleanup goals are achieved (Ref. 21).

> **THE DARAMEND® FACT SHEET PREPARED BY US EPA IS INCLUDED IN APPENDIX C.**

The DARAMEND® technology can be implemented ex situ or in situ. In both cases, the treatment layer is 2 feet (ft) deep, which is the typical depth reached by the specialized deep-tillage equipment. For treatment to greater depths, the technology can be implemented in sequential 2-ft lifts. The DARAMEND® technology may be technically or economically infeasible for excessively high contaminant concentrations in soils (Ref. 21).

DARAMEND® has been used to treat soil and sediment containing low concentrations of pesticides such as toxaphene, HCB, and DDT as well as other contaminants. The technology has not been used for treatment of other POPs such as PCBs, dioxins, or furans.

DARAMEND® has been used to treat POPs at several sites in the US, Canada, Europe and Brazil. In the US, the technology has been implemented at the T.H. Agriculture and Nutrition Superfund site in Montgomery, Alabama, and the W.R. Grace site in Charleston, South Carolina. Table 3-5 presents performance data from these applications. The average treatment cost (in 2004 USD) at the site in Montgomery was $55 per ton; the vendor did not specify the components included in this cost (Refs. 1 and 55). According to the vendor, Adventus Group, costs for in situ treatment have ranged from

> **TECHNOLOGY TYPE: BIODEGRADATION**
>
> **POPS TREATED: TOXAPHENE, HCB, DIELDRIN, DDT, A-HCH, B-HCH AND LINDANE**
>
> **PRETREATMENT: NONE**
>
> **MEDIUM: SOIL AND SLURRY**
>
> **COSTS: $55 PER TON (EX SITU, COST IN 2004 USD) AND $30,000 PER ACRE (IN SITU, FULL-SCALE COST IN 2007)**
>
> *FULL SCALE*
> *EX SITU AND IN SITU*

approximately $40,000 per acre for a pilot study to approximately $30,000 per acre for full-scale applications (in 2007 USD). In addition, the vendor estimates that ex situ treatment costs approximately $55 per ton of soil (Ref. 2).

**Table 3-5. Performance of DARAMEND® Technology**

| Site | Location | Year Implemented | Period (Months) | POP | Quantity of Soil Treated (Tons) | Scale | Untreated Concentration (mg/kg) | Treated Concentration (mg/kg) |
|---|---|---|---|---|---|---|---|---|
| T.H. Agriculture and Nutrition Superfund site | Montgomery, Alabama | 2003 | 5 | Toxaphene | 4,500 | Full | 189 (Mean) | 10 (Mean) |
| | | | | DDT | | | 81 (Mean) | 9 (Mean) |
| W.R. Grace site | Charleston, South Carolina | 1995 | 8 | Toxaphene | 250 | Pilot | 239 | 5.1 |
| | | | | DDT | | | 89.7 | 16.5 |
| Uniroyal Chemical | Ontario, Canada | NA | 9 | DDT | NA | NA | 53.5 | 4.7 |
| Unknown Future Residential Development | Canada | NA | <1 | DDT | 2 acres | Pilot | 2.0 | 0.33 |
| | | | | Dieldrin | | | .064 | .040 |
| ATOFINA Chemicals | Kentucky | NA | <4 | HCB | NA | NA | 10.9 | 1.3 |
| | | | | a-HCH | | | 7,647 | 446 |
| | | | | b-HCH | | | 1,200 | 373 |
| | | | | Lindane | | | 567 | 14 |
| Former obsolete pesticide warehouse | Moldova | NA | <2 | Lindane | NA | Bench | 17 (Mean) | 10 (Mean) |
| Former Agricultural Site | Florida | 2004 | <1 | Dieldrin | 2,600 | Pilot | .046 (Mean) | .015 |
| Hot-Spot Treatment, Former Manufacturing Facility | Southeastern US | 2003 | 8 | Toxaphene | NA | Pilot | 127.7 (Mean) | 8.7 (Mean) |
| Confidential Client | Ontario, Canada | 2007 | <1 | DDT | 147,000 | Full | 2.2 | 0.5 |

Source: Ref. 1
Notes:
DDT = Dichlorodiphenyltrichloroethane        mg/kg = Milligram per kilogram
HCB = Hexachlorobenzene                      NA = Not available
HCH = Hexachlorocyclohexane

DARAMEND® is a proprietary technology provided by Adventus Remediation Technologies, Inc. (ART) in Mississauga, Ontario, Canada. In the US, the technology is provided by ART's sister company, Adventus Americas, Inc. in Bloomingdale, Illinois. Contact information for the technology vendor is provided in Section 5.0.

The technology has been used to treat toxaphene, HCB, dieldrin, DDT, a-HCH, b-HCH and lindane contaminated soil. Based on structural similarity of the POPs treated by DARAMEND® to other POPs described in section 2.6 of this report, this technology can potentially be used to treat other POPs. However, because of the specificity of biochemical reactions, this technology may or may not be effective in treating similar POPs. The most recent application of this technology to treat POPs was in 2007 at a confidential site in Ontario Canada.

### 3.1.4   Gas-Phase Chemical Reduction

Gas-phase chemical reduction (GPCR™) has been used to treat high-strength wastes containing POPs. GPCR™ is an ex situ technology and has been operated in both fixed and transportable configurations.

The technology uses a two-stage process to treat soil contaminated with POPs. In the first stage, contaminated soil is heated in a thermal reduction batch processor in the absence of oxygen to temperatures around 600°C. At high temperature the organic compounds desorb from the solid matrix and enter the gas phase. The treated soil is allowed to cool prior to its appropriate disposal on or off site. In the second stage, the desorbed gaseous-phase contaminants pass to a GPCR™ reactor, where they react with introduced hydrogen gas at temperatures ranging from 850 to 900°C. This reaction converts organic contaminants into primarily methane and water. Acid gases such as hydrogen chloride may also be produced when chlorinated organic contaminants are present. The gases produced in the second stage are scrubbed by caustic scrubber towers to cool the gases, neutralize acids, and remove fine particulates. The off-gas exiting the scrubber is rich in methane and is collected and stored for reuse as fuel. Methane is also used to generate hydrogen for the GPCR™ process in a catalyzed high-temperature reaction. Spent scrubber water is treated by granular activated carbon filters prior to its discharge (Refs. 43 and 44).

| TECHNOLOGY TYPE: THERMAL-CHEMICAL DEGRADATION |
|---|
| POPS TREATED: HCB, DDT, DIELDRIN, PCBS, ALDRIN, DIOXINS, AND FURANS |
| PRETREATMENT: THERMAL DESORPTION |
| MEDIUM: SOIL, SEDIMENT, AND LIQUID WASTE |
| *FULL SCALE* <br> *EX SITU* |

GPCR™ has been implemented at both full and pilot scales to treat solids and liquids contaminated with POPs. The POPs treated include HCB, DDT, dieldrin, aldrin, PCBs, dioxins, and furans. Table 3-6 presents performance information for the technology. In 1992, GPCR™ was field-tested by US EPA's Superfund Innovative Technology Evaluation (SITE) Program to evaluate the performance of the technology at the Bay City Middleground Landfill located in Bay City, Michigan (Ref. 19).

### Table 3-6.  Performance of GPCR™ Technology

| Site | Location | Period | POP | Quantity of Soil Treated | Scale | Destruction Efficiency |
|---|---|---|---|---|---|---|
| Kwinana Commercial Operations | Australia | 1995 to 2000 | PCBs | 2,000 tonnes (2,200 tons) | Full | > 99.9999% |
| | | | DDT | | | > 99.9999% |
| Kwinana Hex Waste Trials | Australia | April 1999 | HCB | 8 tonnes (9 tons) | Full | > 99.9999% |
| General Motors of Canada Limited | Canada | 1996 to 1997 | PCB | 1,000 tonnes (1,100 tons) | Full | > 99.99999% |
| | | | Dioxins | | | > 99.9995% |

Source:  Ref. 43
Notes:
DDT = Dichlorodiphenyltrichloroethane          PCB = Polychlorinated biphenyl
HCB = Hexachlorobenzene

The technology has been selected by the United Nations Industrial Development Organization for a pilot-scale project to treat approximately 1,000 tons of PCB-contaminated waste in Slovakia. The technology has also been licensed in Japan for treatment of PCB- and dioxin-contaminated wastes (Refs. 44 and 71).

GPCR is a thermal-chemical degradation technology that combines high temperature and hydrogen gas to treat POPs. Based on available information, the technology has treated DDT, HCB, PCBs and dioxin in contaminated soil, sediments, and liquids. Due to

> **THE FACT SHEET PREPARED BY IHPA IS AVAILABLE AT** HTTP://WWW.IHPA.INFO/RESOURCES/LIBRARY/.

the high temperature requirement of this technology, GPCR could potentially treat other POPs. This technology was developed by Eco Logic International, Inc. of Ontario, Canada. Bennett Environmental Inc. of Oakville, Ontario, acquired exclusive patent rights to the technology and is currently modifying the technology to improve its cost effectiveness (Ref. 53). The most recent application of this technology was in 2000 to treat PCBs and DDT at a site in Australia. The performance data for this technology was provided by John Vijgen (IHPA). No performance data, process details, or costs for this technology could be obtained directly from the technology vendor. This technology has not been implemented at a full-scale in the US and is currently not commercially available in the US. Contact information for the technology vendor is provided in Section 5.0.

### 3.1.5 Gene Expression Factor

Gene Expression Factor (or simply Factor) is a new technology available to treat soils and sediments contaminated with POPs. The vendor claims that Factor is a site-specific protein that restores the initial protein within the native bacteria species that was either damaged or removed by site contamination. The Factor enhanced bacteria then metabolically transform chlorine attached to hydrocarbon molecules into inert substances. The vendor claims that Factor is non-hazardous and is applied with other soil amendments, such as lime, organic matter (manure, charcoal, etc.), and fertilizers. Once thoroughly

> **TECHNOLOGY TYPE: BIODEGRADATION**
>
> **POPS TREATED: DIELDRIN, DDT, TOXAPHENE, AND PCBS**
>
> **PRETREATMENT: NONE**
>
> **MEDIUM: SOIL AND SEDIMENTS**
>
> *FULL SCALE*
> *EX SITU AND IN SITU*

mixed, the amended soil is irrigated for approximately two months at least twice a day during warm days and every other day during cooler days until confirmation sampling indicates that the chemicals of concern have been removed. In theory, this treatment can be conducted either in situ or ex situ.

According to the vendor, BioTech Restorations, a bench-scale study must be performed on site-specific soil collected from a potential bioremediation site prior to treatment. This study involves sending approximately 3 gallons of soil to BioTech Restorations laboratory for analysis where detailed soil chemistry, biological oxygen demand, biologically available carbon, indigenous bacteria survey and contaminants are analyzed and identified before conducting the bench study. A variety of Factor proteins are then applied to the soil to determine the most cost effective and efficient Factor for that site. The cost for conducting a bench study is $30,000. Actual treatment costs for Factor range from $30 to $60 per ton of contaminated soil, depending on access and site complexity (Ref. 32). Factor has been applied at several sites in California; performance data are included in Table 3-7.

Gene Expression Factor is a bioremediation technology that has been used to treat PCBs, DDT, dieldrin and toxaphene contaminated soil and sediments. Based on structural similarity of the POPs treated by this technology to other POPs described in section 2.6 of this report, Gene Expression Factor may potentially be used to treat other POPs. However, because of the specificity of biochemical reactions involved with this technology, it may or may not be effective in treating similar POPs. This technology was most recently implemented at the Mantegani Property, in South San Francisco, California and was completed in 2007. A fact sheet is not available for this technology. Further technology information can be obtained by contacting the vendor BioTech Restorations. Contact information for the vendor is provided in Section 5.0.

**Table 3-7. Performance of Gene Expression Factor**

| Site, Location | Period | Type | Quantity of Soil Treated (cy) | POP | Initial Concentration (mg/kg) | Final Concentration (mg/kg) |
|---|---|---|---|---|---|---|
| Former North American Transformer South Yard Area, Milpitas, CA | Nov 2005 to Oct 2006 | Ex situ | 15,000 | PCBs | 156 (Max) | <1 |
| Borello Property, Morgan Hill, CA | June 2005 to Aug 2005 | In situ | 14,200 | Dieldrin | 0.48 | 0.00354 (Mean) |
| | | | | Toxaphene | 6.2 | 0.130 |
| Mantegani Property, South San Francisco, CA | May 2005 to Jan 2007 | In situ | 2,200 | DDT | 9.0 (Max) | <0.5 |
| | | | | Dieldrin | 1.0 (Max) | 0.5 |

Source: Ref. 32

Notes:
cy = cubic yard                                          mg/kg = milligrams/kilograms
DDT = Dichlorodiphenyltrichloroethane        PCB = Polychlorinated biphenyl

### 3.1.6    GeoMelt™

**Geomelt ™ ICV process. Ref. 30**

GeoMelt™ has been used to treat high-strength wastes containing POPs. The technology is available for both in situ and ex situ applications and in both fixed and transportable configurations. GeoMelt™ vitrification is a high-temperature technology that uses heat to destroy POPs and to reduce mobility of residual contaminants by incorporating them into a vitrified end product. GeoMelt™'s in situ process is available in two primary configurations: (1) In Situ Vitrification (ISV) and (2) Subsurface Planar Vitrification (SPV™). Both configurations use electrical current to heat, melt, and vitrify material in place. ISV is suitable for treatment to depths exceeding 10 feet below the ground surface. SPV is suitable for more shallow applications. GeoMelt™ also provides a variation of SPV called Deep-SPV, which can vitrify narrow treatment zones deeper than 30 feet.

For treatment, an electric current is passed through soil using an array of electrodes inserted vertically into the surface of the contaminated zone. Because soil is not electrically conductive, a starter pattern of electrically conductive graphite and glass frit is placed in the soil between the electrodes. When power is fed to the electrodes, the graphite and glass frit conduct a current through the soil, heating the area and melting directly adjacent soil. Once molten, the soil becomes conductive. The melting proceeds outward and downward. Typical operating temperatures range from 1,400 to 2,000°C. As the temperature increases, contaminants may begin to volatilize. When sufficiently high temperatures are attained, most organic contaminants are destroyed in situ through thermally mediated chemical reactions, yielding carbon dioxide, water vapor, and sometimes hydrogen chloride gas (if chlorinated contaminants are present). Gaseous reaction products (such as hydrogen chloride) and volatilized contaminants that escape in situ destruction are collected by an off-gas hood and are processed through an aboveground off-gas treatment system before discharge to the atmosphere. When the heating stops, the medium cools to form a crystalline monolith vitrified end product, which encapsulates the nonorganic contaminants that were not destroyed or volatilized (Ref. 40).

> **TECHNOLOGY TYPE: HIGH TEMPERATURE DEGRADATION**
>
> **POPs TREATED: DIELDRIN, CHLORDANE, HEPTACHLOR, DDT, HCB, PCBs, DIOXINS, AND FURANS**
>
> **PRETREATMENT: NONE**
>
> **MEDIUM: SOIL AND SEDIMENTS**
>
> *FULL SCALE*
> *EX SITU AND IN SITU*

GeoMelt[TM]'s ex situ process, which is called In Container Vitrification (ICV[TM]), involves heating contaminated material in a refractory-lined container. A hood placed over the container collects off-gases. The heat is generated by either two or four 12-inch-diameter, graphite electrodes positioned vertically in the container. Typical operating temperatures range from 1,400 to 2,000°C. At these temperatures, the waste matrix melts and organic contaminants are destroyed or volatilized. The off-gas from the process enters an off-gas treatment system, which includes a baghouse particulate filter, high-efficiency particulate air (HEPA) prefiltration, a $NO_x$ (oxides of nitrogen) scrubber, a hydrosonic scrubber, a mist eliminator, a heater, and one or two HEPA filters. After treatment, the hood is removed and a lid is installed on the refractory-lined container. When the melt has solidified, the vitrified waste-filled container is disposed in an appropriate landfill based on the results of US EPA Toxicity Characteristic Leaching Procedure (TCLP) analysis.

> **THE FACT SHEET PREPARED BY IHPA IS AVAILABLE AT HTTP://WWW.IHPA.INFO/RESOURCES/LIBRARY/.**

GeoMelt[TM] is a full-scale thermal degradation technology that uses high temperature (up to 2,000°C) to treat soil and sediments contaminated with POPs such as dieldrin, chlordane, heptachlor, DDT, HCB, PCBs, dioxins, and furans (Ref. 40). GeoMelt[TM] has also been used to treat radioactive waste. Table 3-8 provides performance information for the technology. The use of high temperature destroys and volatilizes the POPs found in contaminated media. The contaminants that are not destroyed are encapsulated in the crystalline monolith vitrified end product. Due to the high temperature requirement of this technology, other POPs could also be potentially treated using GeoMelt[TM]. This technology was originally commercially available from AMEC Earth and Environmental, the sole licensee of this technology in the US. In 2009, IMPACT Services, Inc. (http://www.impactservicesinc.net/ ), a waste processing facility located at the East Tennessee Technology Park, acquired all assets relating to the GeoMelt[TM] business of AMEC Earth and Environmental. The most recent application of this technology to treat POPs in the US was in 2005 at a Wasatch Chemical Superfund. However, GeoMelt[TM] has been extensively used in Japan to treat POP contaminated soil, and it was most recently used to treat POPs in 2008 at MCK facility in Mie Prefecture, Japan. Further information about this technology can be obtained by contacting the vendor (IMPACT Services, Inc.). Vendor contact information is provided in Section 5.0.

**Table 3-8. Performance of GeoMelt™ Technology**

| Site | Location | Period | POP | Quantity of Soil Treated | Scale | Untreated Concentration (mg/kg) | Treated Concentration (mg/kg) |
|---|---|---|---|---|---|---|---|
| Parsons Chemical/ ETM Enterprises Superfund Site | Grand Ledge, Michigan | 1993 to 1994 | DDT | 4,350 tons | Full | 340 (Max) | <0.016 |
| | | | Chlordane | | | 89 (Max) | <0.08 |
| | | | Dieldrin | | | 87 (Max) | <0.016 |
| TSCA Spokane | Spokane, Washington | 1994 to 1996 | PCBs | 5,375 tons | Full | 17,860 | ND |
| Wasatch Chemical | Salt Lake City, Utah | 1995 to 1996 | Dioxins | 5,440 tons | Full | 38 (Max) | ND |
| | | | DDT | | | 1.091 (Max) | ND |
| | | | Chlordane | | | 535 (Max) | ND |
| | | | HCB | | | 17 | <0.08 |
| WCS-Commercial TSCA cleanup | Andrews, Texas | 2005 | PCBs | 5 tons | Full | 496 | ND |
| WCS-Rocky Flats | Andrews, Texas | 2005 | PCBs | 11 tons | Pilot | 130 | ND |
| POPs Agricultural Treatment Project | MCK, Mie Prefecture, Japan | 2006 | Aldrin | 161 tons | 9.5 tonne/ batch | 26 | <0.0003 mg/l |
| | | | HCH | | | 4,000 | <0.0025 mg/l |
| | | | DDT | | | 1100 | <0.0125 mg/l |
| | | | Dieldrin | | | 240 | <0.0003 mg/l |
| | | | Endrin | | | 2 | <0.005 mg/l |
| | | | Dioxins | | | 9.3 ng-TEQ/g | 0.002 pg-TEQ/g |
| Nose Dioxin Contaminated Waste Treatment Project | MCK, Mie Prefecture, Japan | 2006 | Dioxin | 51 tons | 9.5 tonne/ batch | 81 ng-TEQ/g | 0.019 pg-TEQ/g |
| POPs Agricultural Treatment Project | MCK, Mie Prefecture, Japan | 2007 | Aldrin | 209 tons | 9.5 tonne/ batch | 0.0047 mg/l | <0.00000005 mg/l |
| | | | HCH | | | 0.55 mg/l | <0.000007 mg/l |
| | | | DDT | | | 0.094 mg/l | <0.0000022 mg/l |
| | | | Dieldrin | | | 0.0005mg/l | <0.0000002 mg/l |
| | | | Endrin | | | 0.04 mg/l | <0.0000003 mg/l |
| | | | Dioxins | | | 0.0086 ng-TEQ/g | 0.00045 pg-TEQ/g |
| Dioxin Contaminated Sludge Treatment | Nose, Osaka Prefecture, Japan | 2007 | Dioxins | 5.4 tons | 1 tonne/ batch | 1.6 ng-TEQ/g | 0 pg-TEQ/g |

| Site | Location | Period | POP | Quantity of Soil Treated | Scale | Untreated Concentration (mg/kg) | Treated Concentration (mg/kg) |
|------|----------|--------|-----|--------------------------|-------|--------------------------------|-------------------------------|
| POPs Agricultural Treatment Project | MCK, Mie Prefecture, Japan | 2008 | HCH | 71 tons | 9.5 tonne/ batch | 210 | 0.000089 |
| | | | DDT | | | 130 | 0.00011 |

Source:  Ref. 30 and 44

Notes:
DDT = Dichlorodiphenyltrichloroethane
HCB = Hexachlorobenzene
HCH = Hexachlorocyclohexane
mg/kg = Milligram per kilogram
ND = Below detection limit
ng-TEQ/g = Nanogram Toxic Equivalent of Dioxins per gram

PCB = Polychlorinated biphenyl
pg-TEQ/g = Picogram Toxic Equivalent of Dioxins per gram
TSCA = Toxic Substance Control Act
WCS = Wasatch Chemical Superfund

### 3.1.7   Mechanochemical Dehalogenation

Mechanochemical Dehalogenation (MCD™) has been used to treat high-strength wastes containing POPs.  The MCD™ technology uses mechanical energy to promote reductive dehalogenation of contaminants.  In this process, contaminants react with a base metal and a

> **THE FACT SHEET PREPARED BY IHPA IS AVAILABLE AT** HTTP://WWW.IHPA.INFO/RESOURCES/LIBRARY/.

hydrogen donor to generate reduced organics and metal salts.  The base metal is typically an alkali-earth metal, an alkaline-earth metal, aluminum, zinc, or iron.  Hydrogen donors include alcohols, ethers, hydroxides, and hydrides.  The process occurs ex situ in an enclosed ball mill, with a grinding medium to

> **TECHNOLOGY TYPE: PHYSICAL-CHEMICAL DEGRADATION**
>
> **POPS TREATED: DDT, ALDRIN, DIELDRIN, LINDANE AND PCBS**
>
> **MEDIUM:  SOIL, SEDIMENT AND LIQUID WASTES**
>
> **PRETREATMENT: SOIL DRYING AND SCREENING**
>
> *FULL SCALE*
> *EX SITU*

provide mechanical energy and mixing.  The technology is applicable to soil, sediments, and mixed solid-liquid phases.  The by-products generated by the process are reportedly nonhazardous organics and metal salts (Ref. 66).

One MCD™ process developed by Environmental Decontamination Ltd. (EDL) is being used at full scale to treat soil at the Fruitgrowers Chemical Company site in Mapua, New Zealand.  The site is a former pesticide and herbicide manufacturing plant that operated from 1950 to 1980.  The site is approximately 8.4 acres in area and contains about 710,000 cubic feet of soil contaminated with DDT, dichlorodiphenyldichloroethane (DDD),

**MCD process at the Mapua Site.  Ref. 66**

dichlorodiphenyldichloroethylene (DDE), aldrin, dieldrin, and lindane. Proof of performance testing of the MCD™ process was conducted at the site between February 16 and April 23, 2004. The objective of the testing was to demonstrate the technology's ability to treat the contaminated soil to meet cleanup standards for commercial land use. The cleanup criteria are listed in Table 3-9 and the proof of performance testing results are listed in Table 3-10. The criteria are based on the concentration of DDX (the sum of the concentrations of DDT, DDD, and DDE) and the sum of the concentrations of aldrin, dieldrin, and lindane.

**Table 3-9. Soil Acceptance Criteria for the Mapua Site**

| Land Use | Depth (meters) | DDX (Total DDT, DDD, and DDE) (mg/kg) | Aldrin + Dieldrin + Lindane (mg/kg) |
|---|---|---|---|
| Commercial | 0 to 0.5 | 5 | 3 |
| | Below 0.5 | 200 | 60 |

Source: Ref. 66
Notes:
DDD = Dichlorodiphenyldichloroethane          DDX = Total DDD, DDE, and DDT
DDE = Dichlorodiphenyldichloroethylene          mg/kg = Milligram per kilogram
DDT = Dichlorodiphenyltrichloroethane

At the Mapua site, soil with greater than a 10-millimeter (mm) size has contaminant concentrations below the soil acceptance criteria for the site and requires no treatment. EDL receives contaminated soil that is less than 10 mm in size. The 10 mm size fraction is dried and passed through a 2-mm screen to segregate soil particles less than and greater than 2 mm size. Contaminated soil less than 2-mm size is treated using the MCD™ process. Additional information on the soil drying, soil screening, and MCD™ processing are described below.

Soil Drying: Contaminated soil with a size of 10 mm or less enters a temperature controlled, diesel-fired rotary drum unit. As the soil passes through the dryer, the soil particles undergo size reduction. The moisture content in soil exiting the dryer is typically less than 2 percent. Gaseous emissions from the dryer are treated by an air quality control system consisting of cyclones, a baghouse, a scrubber and an activated carbon filter.

Soil Screening: Soil exiting the dryer is passed through a rotary screen to separate soil particles by size. Soil particles less than 2 mm in size are separated from soil particles between 2 and 10 mm in size. Soil samples are collected from the 2- to 10-mm fraction stream and analyzed. Thus far, DDX concentrations in this size fraction have been at or below cleanup standards and have consequently not required treatment (Ref. 66). The less than 2-mm size fraction and the fines from the cyclones and baghouse are fed into the MCD™ reactor.

MCD™ Processing: Dried contaminated soil (the less than 2-mm fraction) and fines from the cyclones and baghouse are fed into the MCD™ reactor and are mixed with metered quantities of a combination of metal salts and a hydrogen donor at a rate of around 3 percent by mass. The reactor is a vibratory mill with two horizontally mounted cylinders containing a grinding medium. The grinding medium provides the mechanical impact energy required to drive the chemical reaction. Treated soil exits the base of the MCD™ reactor through enclosed screw conveyors and enters a paddle mixer, where the treated material is wetted to minimize dust generation. The required residence time within the reactor is about 15 minutes. The treated soil is then analyzed. Once treatment of soil to cleanup standards has been completed, treated soil is placed in a clean backfill area.

During proof of performance testing at the Fruitgrowers Chemical Company site in Mapua, New Zealand, the MCD™ system treated a maximum of 139 cubic meters per week. Table 3-10 lists the initial and final contaminant concentrations in the soil treated by the MCD™ reactor. The concentrations listed in Table 3-10 are mean concentrations in samples collected. The treated soil met the cleanup criteria for soil taken from a depth of over 0.5 meter (m) below ground surface, but did not meet the criteria for surface soil taken from between 0 and 0.5 m below ground surface.

**Table 3-10. Performance of MCD™ Technology at the Mapua Site**

| POP | Untreated Concentration (mg/kg) | Treated Concentration (mg/kg) | Percent Reduction | Soil Acceptance Criteria (mg/kg) by Depth Below Ground Surface | |
|---|---|---|---|---|---|
| | | | | 0 to 0.5 meters | > 0.5 meters |
| DDX | 717 (Mean) | 64.8 (Mean) | 91% | 5 | 200 |
| Aldrin | 7.52 (Mean) | 0.798 (Mean) | 89% | NA | NA |
| Dieldrin | 65.6 (Mean) | 19.8 (Mean) | 70% | NA | NA |
| Lindane | 1.25 (Mean) | 0.145 (Mean) | 88% | NA | NA |
| Aldrin+Dieldrin +Lindane | 73.245 (Mean) | 20.612 (Mean) | 72% | 3 | 60 |

Source: Ref. 66
Notes:
DDX = Total Dichlorodiphenyldichloroethane (DDD), Dichlorodiphenyldichloroethylene (DDE), and
    Dichlorodiphenyltrichloroethane (DDT)
mg/kg = Milligram per kilogram
NA = Not available

Subsequent to the proof of performance testing, EDL began a full-scale application later in 2004. By early December 2006, a total soil/sediment volume of 55,250 cubic meters was excavated, screened, relocated, or treated. Of this volume, approximately 5,500 cubic meters were treated using MCD™. Cleanup completion was scheduled for March 2007 with a total project cost of $8 million, including construction and continuous operation for 2.5 years (Ref. 28).

EDL has conducted two treatability studies at Hunters Point Shipyard in San Francisco, California. The initial study was conducted in 2006 to evaluate the feasibility of remediating PCB contaminated soils. A supplemental study was also conducted from August to November 2007 to evaluate treatment of additional on-site soils. Initial PCB concentrations of approximately 300 mg/kg were reduced to less than 1 mg/kg during the first and supplemental studies. The supplemental study also evaluated post-treatment stabilization of soils to (1) stabilize heavy metals in the soil matrix and (2) restore the physical characteristics of the treated soil converting it from a powdery form to a texture similar to garden soil. Based on results from these studies, a full-scale MCD™ system with a capacity of 10 metric tons per hour could be used at this site (Ref. 57). The performance data for this application could not be obtained from the technology vendor.

MCD™ uses mechanical energy in combination with a base metal and a hydrogen donor to promote dehalogenation of POPs. This technology has been used to treat DDT, aldrin, dieldrin and lindane. Based on structural similarity of known POPs treated using MCD to other similar POPs described in section 2.6 of this report, this technology can potentially be used to treat other POPs. However, the potential of MCD™ technology to treat other POPs is highly dependent on the base metals and hydrogen donors used and the specific chemical reactions that occur under the treatment conditions. The most recent application of this technology to treat POPs in the US was in 2005 at Hunters Point Shipyard in San Francisco, California. The technology was also implemented in 2004 to treat POPs contaminated soil at the Fruitgrowers Chemical Company site in Mapua, New Zealand. The MCD™ technology is available from

EDL in Auckland, New Zealand (http://www.edl-asia.com/home.html) and from Tribochem in Wunstrof, Germany (http://www.tribochem.com) (Ref. 13). Information for this section of the report was provided by EDL. Contact information for EDL is provided in Section 5.0. Tribochem has not provided process details, performance data, or costs for its technology. Currently, no vendor is available in the US for this technology.

### 3.1.8   Plasma Arc

Plasma arc technologies use a thermal plasma field to treat contaminated wastes. The plasma field is created by directing electric current through a gas stream under low pressure to form a plasma with a temperature ranging from 1,600 to 20,000°C. Bringing the plasma into contact with the waste causes contaminants to dissociate into their atomic elements. The separated elements are subsequently cooled, which causes them to recombine to form inert compounds. The process may also destroy organic compounds through pyrolysis. The end products are typically gases, such as carbon monoxide, carbon dioxide, hydrogen and inert solids. If chlorinated compounds are present in the waste, acid gas is also generated

> TECHNOLOGY TYPE: THERMAL DEGRADATION
>
> POPs TREATED: PCBS, CHLORDANE, DDT, ENDOSULFAN, DIOXINS, AND FURANS
>
> MEDIUM: SOLID AND LIQUID WASTES
>
> PRETREATMENT: THERMAL DESORPTION
>
> *FULL SCALE*
> *EX SITU*

as an end product. The off-gas from the plasma arc system passes through an off-gas treatment system and is then discharged. The plasma arc technologies that are used to treat organic wastes include PLASCON™, Plasma Arc Centrifugal Treatment (PACT™), and the Plasma Converter System (PCS).

This report focuses on PLASCON™, which has been used at full scale to treat POPs. PLASCON™ is an ex situ technology that can treat both solid and liquid waste streams. It is potentially applicable to both low- and high-strength wastes containing POP contamination. The PLASCON™ technology passes a direct current discharge through argon gas to create plasma with a temperature greater than 10,000°C. Liquid or gaseous waste is injected directly into the plasma. Solid waste is pretreated using thermal desorption to extract volatile contaminants. The extracted vapors are then condensed and injected into the plasma as liquid waste. Liquid waste is vaporized by heat transfer from the plasma. Organic compounds present in the waste pyrolize. The products formed during pyrolysis pass through a reaction tube providing sufficient residence time to ensure complete decomposition of the feed material. Gases exit the tube at a temperature of about 1,500°C and are rapidly cooled to less than 100°C in a spray condenser

> THE FACT SHEET PREPARED BY IHPA IS AVAILABLE AT
> HTTP://WWW.IHPA.INFO/RESOURCES/LIBRARY/.

using an alkaline spray solution. The gases are further cooled and scrubbed of any remaining acid gases in a packed tower. Off-gases, which contain mainly carbon monoxide and argon, are then thermally oxidized to convert carbon monoxide to carbon dioxide (Ref. 44).

Several full-scale applications of this technology have been performed to treat POPs around the world. At least nine commercial plants are in operation (four in Japan, four in Australia, and one in the United Kingdom). One plant in Japan operated by the Mitsubishi Chemical Corporation treated more than 1,000 tons of PCB contaminated waste from May through July 2004 (Ref. 42). A facility in Brisbane, Australia treats concentrated PCB solutions (>10 %) as well as a range of POP pesticides. Information about these applications was provided by John Vijgen (IHPA). No performance data could be obtained directly from the technology vendor.

PLASCON™ uses high temperatures and heated plasma to degrade POP contaminants found in soil and liquid waste into their individual atomic elements. This technology has been used to treat various POPs such as PCBs, DDT, chlordane, endosulfan, dioxins and furans. Due to the high temperature requirement of this technology (temperature ranging from 1,600 to 20,000°C), other POPs could also be potentially treated using PLASCON™. SRL Plasma Pty. Ltd., an Australian company, is the patent holder of this technology. The most recent application of this technology to treat POPs was in 2004 in Japan. Currently, no know vendors for this technology exist in the US. Contact information for SRL Plasma Pty. Ltd. is provided in Section 5.0.

### 3.1.9 Radicalplanet® Technology

Radicalplanet® technology is an ex situ process that uses mechanochemical principle to treat POPs (Ref. 51). It has been used to treat PCBs, chlordane, DDT, endrin, dioxins, and furans. This process transforms POPs molecules into their "radical" state by use of the "planetary mill." The treatment occurs in a reaction vessel where steel balls and a reagent chemical, such as calcium oxide (CaO), are placed prior to the placement of the wastes. The vessels are then sealed and placed on the Radicalplanet® machine where the vessels are rotated (rotation speed: 70 to 100 revolutions per minute [rpm]). As the steel balls crash into each other, the bonds of the POPs are broken by the mechanical energy. At a rotation speed of 100 rpm, the dechlorination reaction is complete in about three to six hours.

> **TECHNOLOGY TYPE: PHYSICAL-CHEMICAL DEGRADATION**
>
> **POPS TREATED: PCBS, CHLORDANE, DDT, ENDRIN, DIOXINS, AND FURANS**
>
> **MEDIUM: SOLID WASTE, SOIL, ASH**
>
> **PRETREATMENT: NONE**
>
> *FULL SCALE*
> *EX SITU*

The reaction vessel is cooled externally by circulating cooling water. The treatment technology does not require pretreatment and does not require a high amount of power to operate. No effluent or off-gases are generated from this treatment process. The reaction vessels are mobile and are available in two different sizes; an E-200 Type reaction vessel that can hold up to 750 liters of contaminated material and an A-500 Type reaction vessel that can hold up to 1500 liters of contaminated material.

**Radicalplanet® Technology Ref. 51**

Radicalplanet® technology uses mechanical energy in the presence of chemicals like CaO to promote dehalogenation of contaminants. This technology has been used to treat BHC, DDT, endrin, HCH, PCBs and Dioxin. Based on structural similarity of known POPs treated using Radicalplanet® to other similar POPs described in section 2.6 of this report, this technology can potentially be used to treat other POPs. However, the potential of Radicalplanet® technology to treat other POPs is highly dependent on the chemicals used and the specific chemical reactions that occur under the treatment conditions. According to the technology's vendor, Radicalplanet® technology was evaluated and approved by three Japanese Ministries of Government. Currently, this technology is available only in Japan and further information can be obtained from the technology's vendor, Radicalplanet® Research Institute Co. Ltd. Performance data is provided in Table 3-11. Vendor contact information is provided in Section 5.0.

**Table 3-11. Performance of Radicalplanet® Technology at Various Japanese Sites**

| Contaminant(s) and Medium | Site Location | Untreated Concentration (mg/kg) | Treated Concentration (mg/kg) | Percent Reduction | Treated Dioxin Toxic Equivalent (pg-TEQ/g) |
|---|---|---|---|---|---|
| BHC (powder) | Ibraki, Japan (with Institute of Environmental Toxicology) | 970,000 | 0.16 | 99.9999% | <1 |
| DDT (powder) | | 50,000 | 0.001 | 99.9999% | <1 |
| Endrin (powder) | | 20,000 | 0.01 | 99.9999% | <1 |
| HCH (powder) | | 50,000 | 0.01 | 99.9999% | <1 |
| PCB (soil) | Ibraki, Japan | 42,800 | 0.01 | 99.9999% | <1 |
| PCB (soil) | Ibraki, Japan (with Geo-Environmental Protection Center) | 75,000 ng-TEQ | 0.13 ng-TEQ | 99.9999% | <1 |
| Dioxin (ash) | | 81,000 ng-TEQ | 0.15 ng-TEQ | 99.9999% | <1 |

Source: Ref. 58
Notes:
BHC = Hexachlorobenzene
DDT = Dichlorodiphenyltrichloroethane
HCH = Hexachlorocyclohexane
PCB = Polychlorinated biphenyl

pg-TEQ/g = Picogram Toxic Equivalent of Dioxins per gram
ng-TEQ/g = Nanogram Toxic Equivalent of Dioxins per gram
mg/kg = Milligram per kilogram

### 3.1.10 Solvated Electron Technology

Solvated Electron Technology (SET™) is a non-thermal chemical degradation treatment process that has been used to treat PCB contaminated soil. The SET™ process occurs in a closed system and uses solvated electron solutions to reduce organic compounds to metal salts and the parent dehalogenated molecules. Solvated electron solutions are formed by dissolving alkali or alkaline earth metals such as sodium, calcium, and lithium in solvents such as anhydrous liquid ammonia. The solvated electron solution is added to the treatment cell containing the waste and ammonia solution. A chemical reaction occurs in the treatment cell between the solvated electrons and the contaminants instantly. After the reaction is complete, hot water or steam is circulated through the jacket of the treatment cell. The warmed ammonia is then removed from the treatment cell and recovered for reuse through the ammonia recovery system. At the end of the reactions, the treated material left behind in the treatment cell may have a high pH and is adjusted using a dilute acid solution prior to disposal. According to the vendor of this technology, the technology does not produce regulated by-products such as dioxins or furans or their precursors.

> **TECHNOLOGY TYPE: CHEMICAL DEGRADATION**
>
> **POPS TREATED: PCBS**
>
> **MEDIUM: SOIL**
>
> **PRETREATMENT: NONE**
>
> *FULL SCALE*
> *EX SITU*

SET™ has been used to treat PCB contaminated oil, mixed waste and soil. It was used at full-scale to treat PCB contaminated soil at the Pennsylvania Air National Guard Site in Harrisburg, Pennsylvania. The site was contaminated with PCBs from an electrical transformer dielectric spill that occurred in 1979. PCB concentrations in the soil ranged from 17 to 560 ppm. Approximately, 340 tons of soil was excavated from the spill area and treated using the SET™ process from May 2000 to July 2001. Post treatment sampling indicated that the PCB concentrations were less than 1 ppm and the treated soil was used to backfill the excavated area (Ref. 33).

SET™ is a chemical degradation process that uses solvated electron solutions to treat POP-contaminated soils. Commodore Advanced Sciences, Inc., based in Washington State, developed and holds the patent for this technology. This technology has been used to treat PCBs. Based on structural similarity of PCBs to other similar POPs described in section 2.6 of this report, this technology can potentially be used to treat other POPs. However, the potential of SET™ to treat other POPs is dependent on the solvated electron solution used and the specific reactions that occur under the treatment conditions of the technology. The most recent application of this technology to treat POPs, PCB-contaminated soil, was in 2001 at the Pennsylvania Air National Guard Site in Harrisburg, Pennsylvania. No fact sheet is available for this technology. Further information about this technology can be obtained by contacting the vendor (Commodore Advanced Sciences, Inc). Vendor contact information is provided in Section 5.0.

**Typical SET™ Plant**

### 3.1.11 Sonic Technology

Sonic Technology is an ex situ process used to treat low- and high-strength soils containing PCB contamination. Pretreatment is performed using the Terra-Kleen process, which involves mixing contaminated soil with a solvent to produce a slurry. The solvent extracts the contaminants from the soil and concentrates them in a residual waste stream, which is collected in a receiving tank. After the extraction is completed, the waste stream with PCB concentrate is diluted, mixed with proprietary method for creating a self-regenerated sodium dispersion chemical destruction and subjected to the sonic energy generated by a proprietary low-frequency generator (Sonoprocess™). The sonic energy activates dechlorination of the PCBs in the solvent. The spent solvent can then be recycled through the system. Any off-gas from the process is treated using condensation, demisting, and multistage carbon filtration (Ref. 59).

| TECHNOLOGY TYPE: PHYISCAL-CHEMICAL DEGRADATION |
| :---: |
| POPS TREATED: PCBS |
| MEDIUM: SOIL |
| PRETREATMENT: TERRA-KLEEN |
| *FULL SCALE EX SITU* |

In 2006, the technology was implemented at full scale to treat approximately 3,000 tons of PCB-contaminated soil at the Juker Holdings site in Vancouver, British Columbia, Canada. At this site, PCB concentrations in soil were reduced from an initial range of 400 to 1,600 ppm to <25 ppm. The modified Terra-Kleen solvent extraction process isolated the PCB fraction in an oil matrix. The final concentration of the oil contained 46,000 ppm of PCB at the start of the application of the sonic, low frequency sonicator process. The process uses ordinary solid sodium ingots and creates an in situ, self-generating sodium dispersion that is much lower in cost and more effective than purchasing pre-made sodium dispersions. The final treatment process treated approximately 11,000 gallons of PCB concentrate to <3 mg/kg within two weeks. This project was completed in its entirety in 2007. The equipment is skid-mounted and transportable.

**Sonic Technology process. Ref. 59**

Sonic Technology is a physical-chemical degradation process that uses solvent extraction process as a pre-treatment stage to extract the contaminants in solution and uses low frequency sonic energy to dehalogenate chlorinated compounds. This technology has been used to treat PCB-contaminated soil. The potential of Sonic Technology to treat other POPs is dependent on the proprietary solvent used to extract the POPs in solution and the frequency of sonic energy generated by the Sonoprocess™. The most recent application of this technology to treat POPs was in 2006 at Juker Holdings site in Canada. Vendor contact information is provided in Section 5.0.

### 3.1.12 Thermal Desorption

Thermal desorption is a physical separation process which heats wastes to volatilize water and organic contaminants. A vacuum system or a carrier gas transports the volatilized organic contaminants to an off-gas treatment system. In the off-gas treatment system, particulates, if present, are removed by conventional particulate removal equipment (such as wet scrubbers or fabric filters) and contaminants are removed either through condensation followed by carbon adsorption, or they are destroyed in a secondary combustion chamber or a catalytic oxidizer. Three types of conventional mobile or fixed thermal desorption units are available, including: direct fire, indirect fire, and indirect heat. In the direct fire type, fire is applied directly upon the surface of contaminated media to desorb contaminants from the soil. In the indirect fire type, a direct-fired rotary dryer heats an air stream which, by direct contact, desorbs water and organic contaminants from the soil. In the indirect heat type, an externally fired rotary dryer volatilizes the water and organics from the contaminated media into an inert carrier gas stream. The carrier gas is later treated to remove or recover the contaminants. Based on the operating temperature of the desorber, conventional thermal desorption processes can be categorized into two groups: high temperature thermal desorption (HTTD) and low temperature thermal desorption (LTTD). HTTD is a full-scale technology in which wastes are heated to 320 to 560°C (600 to 1,000°F). In LTTD, wastes are heated to between 90 and 320°C (200 to 600°F).

**ISTD process at the Alhambra site. Ref. 63**

This report focuses on in situ thermal desorption (ISTD), which has been used to

**THE FACT SHEET PREPARED BY US EPA IS INCLUDED IN APPENDIX D.**

treat both high- and low-strength wastes containing POPs. ISTD is primarily an in situ technology but has also been used ex situ on constructed soil piles. ISTD is a thermally enhanced, in situ treatment technology that uses conductive heating to transfer heat directly to environmental media. Although there are other in-situ thermal technologies commercially available, ISTD is the only in situ technology that can reach the elevated temperatures (greater than 100°C) required for treatment of low volatility POPs such as dioxins and DDT. ISTD, sometimes also known as "In Situ Thermal Destruction," is a patented technology developed by Shell Oil Co., and the patent was donated to The University of Texas at Austin. TerraTherm holds the exclusive license to the technology in the United States and is the only vendor. TerraTherm partners with other companies abroad to apply the technology internationally.

The ISTD process includes three basic elements (Ref. 63):

1. Application of heat to contaminated media by thermal conduction
2. Collection of desorbed contaminants through vapor extraction
3. Treatment of collected vapors

In the most common setup, ISTD uses a vertical array of electrically powered heaters placed in wells drilled into the remediation zone. The heaters reach temperatures in excess of 600°C and heat contaminated media by thermal conduction. Surface heating blankets or in pile desorption units are less commonly used. As the matrix is heated, adsorbed and liquid-phase contaminants begin to vaporize. For treatment of POPs, after target soil temperatures (typically 335°C) are achieved, a portion of the organic contaminants either oxidizes (if sufficient air is present) or pyrolizes. In order to reach the treatment temperatures required for the least volatile POPs, the treatment must be above the water table or the influx of water must be significantly reduced.

A network of vapor extraction wells is used to recover volatilized contaminants. Contaminant vapors captured by the extraction wells are conveyed to an off-gas treatment system for treatment before discharge to the atmosphere. TerraTherm offers two different methods of vapor treatment. One method treats extracted vapor without phase separation and uses a thermal oxidizer to break down organic vapors to primarily carbon dioxide and water. Thermal oxidation may be followed by vapor phase activated carbon adsorption. The second method uses a heat exchanger to cool extracted vapors. The resulting liquid phase is then separated into aqueous and nonaqueous phases. The nonaqueous-phase liquid (NAPL) is usually disposed of at a licensed treatment, storage, or disposal facility. The aqueous phase is

> TECHNOLOGY TYPE: THERMAL REMEDIATION
>
> POPS TREATED: PCBS, DIOXINS, AND FURANS
>
> MEDIUM: SOIL AND SEDIMENT
>
> PRETREATMENT: NONE
>
> COSTS: $200 TO $600 PER CUBIC YARD (COST IN USD, DATA FROM 1996 TO 2009)
>
> *FULL SCALE*
> *IN SITU*

passed through liquid-phase activated carbon adsorption units and is then discharged. Cooled, uncondensed vapor is passed through vapor-phase activated carbon adsorption units and is then vented to the atmosphere (Refs. 10, 26 and 63).

Pilot- and full-scale applications of ISTD have been used to address PCBs, dioxins, and furans. According to TerraTherm, laboratory-scale work indicates that this technology can also effectively treat other POPs, including aldrin, dieldrin, endrin, chlordane, heptachlor, DDT, mirex, HCB, and toxaphene. However, these contaminants have not yet been treated using ISTD at pilot or full scale. ISTD was field-tested by US EPA's SITE Program to evaluate the performance of the technology at the Rocky Mountain Arsenal site near Denver, Colorado. The site was contaminated with hexachlorocyclopentadiene, aldrin, chlordane, dieldrin, endrin, and isodrin. After 12 days of operation, the ISTD system was shut down because portions of the aboveground piping had been corroded by hydrochloric acid that was generated during the heating of the contaminants. Shutdown of the system prevented the evaluation of the

43

effectiveness of the technology at this site (Ref. 25). Thus, highly concentrated chlorinated wastes that can decompose at the temperatures required for treatment may require expensive materials to reduce corrosion.

Four full-scale and two pilot-scale ISTD projects at POP-contaminated sites were identified. In general, treatment costs in USD at these sites ranged from $200 to $600 per cubic yard (cy). Projects involving ISTD treatment of larger volumes of waste may have lower unit costs. Available performance information for the technology is presented in Table 3-12.

**Table 3-12. Performance of ISTD Technology**

| Site | Location | Period | POP | Quantity of Soil Treated (cy) | Scale | Untreated Concentration (mg/kg) | Treated Concentration (mg/kg) |
|------|----------|--------|-----|-------------------------------|-------|----------------------------------|-------------------------------|
| Former South Glens Falls Dragstrip | Moreau, New York | 1996 | PCBs | NA | Full | 5,000 (Max) | 0.8 |
| Tanapag Village | Saipan, Northern Mariana Islands | July 1997 to August 1998 | PCBs | 1,000 | Full | 500 (Mean) | <10 |
| Centerville Beach | Ferndale, California | September to December 1998 | PCBs | 667 | Full | 0.15-860 | <0.17 |
| | | | Dioxins and Furans | | | 3.2 μg/kg (Max) | 0.006 μg/kg |
| Missouri Electric Works | Cape Girardeau, Missouri | March to June 1997 | PCBs | NA | Pilot | 20,000 (Max) | <0.033 |
| | | | Dioxins and Furans | | | 6.5 μg/kg | .003 μg/kg |
| Former Mare Island Naval Shipyard | Vallejo, California | September to December 1997 | PCBs | 222 | Pilot | 2,200 (Max) | <0.033 |
| Alhambra "Wood Treater" | Alhambra, California | May 2002 to September 2005 | Dioxins and Furans | 16,200 | Full | 19.4 μg/kg (Max) | <.11 μg/kg |

Source: Refs. 9, 11, 38, 60, 62 and 63

Notes:
cy = Cubic yard
μg/kg = Microgram per kilogram
mg/kg = Milligram per kilogram

NA = Not available
PCB = Polychlorinated biphenyl

ISTD is a physical separation process that uses high temperatures to desorb and volatilize contaminants. Some recovered contaminants may oxidize or pyrolize, during the process. The contaminant vapors are then captured and treated using an off-gas treatment system. ISTD has been used to treat POPs such as PCBs, dioxins and furans. Due to the high temperature requirement of this technology, other POPs could also be potentially treated using ISTD. The most recent full-scale application of this technology to treat POPs was at the Alhambra "Wood Treater" Site in California and was completed in 2005. In 2009, TerraTherm and SheGoTec Japan completed a demonstration under the sponsorship of the Japan Ministry of the Environment on dioxin-contaminated soils. The remedial goal of less than 1,000 pg-TEQ/g was achieved while meeting all vapor emission standards in Japan. Further information about this technology can be obtained by contacting the vendor (TerraTherm; http://www.terratherm.com/). Vendor contact information is provided in Section 5.0.

## 3.2    Pilot-Scale Technologies for Treatment of POPs

This section describes technologies that have been implemented to treat POPs at the pilot scale. Each subsection focuses on a single technology and includes a description of the technology and information about its application at specific sites. Fact sheets developed by IHPA provide additional details on some of these technologies and their applications. Links to the IHPA fact sheets are included in the appropriate subsections of this report.

### 3.2.1    Phytotechnology

Phytotechnology is a process that uses plants to remove, transfer, stabilize, or destroy contaminants in soil, sediment, and groundwater. It may be applied in situ or ex situ to treat low-strength soils, sludges, and sediments contaminated with POPs. The mechanisms include:

> TECHNOLOGY TYPE:
> PHYTOREMEDIATION
>
> POPs: DDT, CHLORDANE, AND PCBs
>
> MEDIUM:  SOIL AND SEDIMENT
>
> PRETREATMENT: NONE
>
> *PILOT SCALE*
> *EX SITU AND IN SITU*

- Enhanced rhizosphere biodegradation (degradation in the soil immediately surrounding plant roots),
- Phytovolatilization (the transfer of the pollutants to air via the plant transpiration stream),
- Phytoextraction (also known as phytoaccumulation, the uptake of contaminants by plant roots and the translocation/accumulation of contaminants into plant shoots and leaves),
- Phytodegradation (metabolism of contaminants within plant tissues),
- Phytostabilization (production of chemical compounds by plants to immobilize contaminants at the interface of roots and soil),
- Hydraulic control (the use of trees to intercept and transpire large quantities of groundwater or surface water for plume control), and
- Evapotranspiration (the use of the ability of plants to intercept rain to prevent infiltration and take up and remove significant volumes of water after it has entered the subsurface to minimize the percolation into the contained waste).

In general, more data are available for field studies that have been conducted using phytostabilization and hydraulic control mechanisms. Other proven uses of phytotechnologies include use as alternative landfill caps, the use of wetlands to improve water quality, and treatment of certain contaminants (such as petroleum products and chlorinated solvents).

Phytoremediation of POPs is not feasible for highly contaminated soil, since high concentrations of POPs are toxic to plants, but it can be used as an appropriate polishing technology for residual contamination in soils. Initial laboratory research identified enhanced degradation of PCBs in the rhizosphere (Refs. 17, 34, and 47). Other researchers are finding promising results for phytoextraction in the laboratory and at the pilot scale. The Connecticut Agricultural Experimental Station's preliminary data have shown that a narrow range of plant species (certain cucurbitas) can effectively accumulate significant amounts of highly weathered pesticide residues such as DDE and chlordane from soil (Ref. 73).

In 2002, the Royal Military College (RMC) of Canada performed a three-part study to evaluate the treatment of PCB contaminated soil by phytotechnology (Ref. 75). Initially, a Greenhouse Treatability Study was conducted to determine the uptake of PCBs by pumpkin (*Cucurbita pepo cv. Howden*), tall fescue (*Festuca Arundinacea*) and Sedge. Initial PCBs concentrations in soil used for the greenhouse study ranged from 27.5 to 3050 ug/g. Plant uptake for the treatability study is provided in Table 3.13. Based on the success of the treatability study, a field study was conducted using larger containers and no coverings. The results of this study indicated an increased uptake of PCBs by all three species. The results of the field experiment showed that a significant amount of PCBs were not released into the environment through volatilization. In 2004, a full-scale application was started at a site in Etobicoke, Ontario. The soil was contaminated with PCBs at an average concentration of 21 ppm. The PCB concentrations in the plant stem and leaf in the second season were higher (11 and 8.9 ppm, respectively) than the concentrations in the first season (5.7 and 3.9 ppm, respectively). Consistent uptake has been observed, however, no difference was noticed in the soil PCB concentration before and after the field study took place (Ref. 74).

**Table 3-13. Results of Plant Uptake from the Royal Military College Study**

|  | Greenhouse Treatability Study | | Small Field Experiment | |
|---|---|---|---|---|
|  | Roots (µg/g) | Shoots(µg/g) | Roots(µg/g) | Shoots(µg/g) |
| **Pumpkin** | 730 | 16.8 | 790 | 370 |
| **Fescue** | 440 | 6.2 | 805 | 580 |
| **Sedge** | 1200 | 470 | 785 | 410 |

Source: Ref. 75
Note: µg/g = Microgram per gram

Research has also been conducted in Ukraine and Kazakhstan on the use of plants to remediate soils laced with pesticides. In the Ukraine, laboratory experiments have shown that bean plants can accumulate and decompose DDT (Ref. 50). In Kazakhstan, native vegetation that can tolerate and accumulate pesticides has been identified (Ref. 52). While research is still active and needed, field-scale projects are also being implemented. A cleanup project was conducted at a 40-year-old scrap yard site with PCB contaminated soils at 225 ppm. The site contamination was approximately 2 acres in area and 3 feet deep. The project demonstrated that PCB concentrations decreased over 90% within 2 years, in the presence of red mulberry trees and bermuda grasses (Ref. 39).

In 2001, phytotechnology was demonstrated by US EPA's SITE Program to evaluate the performance of the technology at the Jones Island Confined Disposal Facility Site in Milwaukee, Wisconsin. In this demonstration, four different treatments were evaluated for treating PCB contaminated soil. Three plant-based (corn, willow and natural vegetation) and one microbe-based treatment were planted at the site. The initial PCB soil concentrations in the test plots ranged from 2.0 to 3.6 mg/kg. At the end of the demonstration in September 2002, the final results indicated that none of the treatments produced a final mean concentration of total PCBs below the cleanup standard of 1 ppm. Further information about the Jones Island Confined Disposal Facility Site can be found at http://www.epa.gov/nrmrl/pubs/540r04508/540r04508.htm.

Furthermore, two US EPA Superfund sites have utilized phytotechnology as a treatment for POPs:

- Aberdeen Pesticides Dumps in North Carolina is using phytotechnology for residual contaminants (dieldrin, DDT, HCB and HCH) (plantings of poplar trees and grasses over 7.5 acres). This technology was selected in 2003 and continues at this site. Updated information is not available for this project.
- Fort Wainwright in Alaska used ex situ phytotechnology for aldrin, dieldrin and DDT with Felt Leaf Willow trees. After treatment, the soil was deposited in the site landfill. This project was completed in 2002.

In general, the science of using phytotechnologies to remediate PCB soil has not had major advancements since the previous edition of this report in 2005. Researchers are still working with *Cucurbita pepo* species and looking at ways such as fertilizer and manipulation of vegetation to enhance the biomass of the plants.

> **FURTHER INFORMATION ABOUT PHYTOTECHNOLOGY CAN BE FOUND AT** HTTP://WWW.CLU-IN.ORG/TECHFOCUS/DEFAULT.FOCUS/SEC/PHYTOTECHNOLOGIES/CAT/OVERVIEW/

### 3.2.2 Reductive Heating and Sodium Dispersion

Reductive heating and sodium dispersion is an ex situ thermo-chemical technology for treating POPs. In the first part of the process, POP-contaminated wastes are indirectly heated using a reductive heating kiln at temperatures ranging from 350 to 600°C in an oxygen-controlled atmosphere, and POPs are decomposed and evaporated from the wastes. The decomposed and evaporated POPs are collected via the oil scrubber, and at the same time evaporated water from the wastes is also condensed. The decomposed and evaporated POPs are dissolved in an oil phase, and the condensed water is accumulated at the bottom of the scrubber. The scrubbing oil containing dissolved POPs and their decomposed substance is treated by a batch method using metallic sodium powder dispersion at a temperature of about 90°C for one hour. After the reaction, water is added to remove excess sodium and settled to separate the treated oil and alkali solution. Separated, treated oil is recycled to serve as the scrubbing oil (Ref. 44).

> **TECHNOLOGY TYPE: THERMAL-CHEMICAL DEGRADATION**
>
> **POPs: DDT, CHLORDANE, ALDRIN, PCBS, B-HCH, DIOXINS AND FURANS**
>
> **MEDIUM: SOIL**
> **PRETREATMENT: NONE**
>
> *PILOT SCALE*
> *EX SITU*

> **THE FACT SHEET PREPARED BY IHPA IS AVAILABLE AT** HTTP://WWW.IHPA.INFO/RESOURCES/LIBRARY/

In a pilot-scale demonstration in Okinawa, Japan, a mobile system was used to treat soil contaminated with BHC and PCBs. The reductive heating kiln had a diameter and length of 400 mm each. The system operated at a maximum capacity of 500 kg/day and each batch was heated at 600°C in an oxygen deficient atmosphere for 1 hour. The exhaust gas from the kiln was passed through a scrubber. The POPs retained were condensed in oil and then treated using sodium dispersion at 90°C for 1 hour. BHC was reduced from 10 mg/kg to less than 0.001 mg/kg in each type of soil, and PCBs were reduced from 53 mg/kg to less than 0.5 ug/kg by reductive heating. The treated soil was then tested for its use as a recycled planting material, and results confirmed its applicability (Ref. 45). Information about this application was provided by John Vijgen (IHPA).

Reductive heating and sodium dispersion combines thermal degradation with sodium dispersion to treat POPs. The initial heating process removes POPs from the contaminated soil followed by the use of metallic sodium to decompose the POP contaminant. This technology has been used at a pilot-scale to treat BHC and PCBs. Due to the high temperature requirement of this technology in the reductive heating process, other POPs could also be potentially treated using this technology. Powertech Labs Inc. from British Columbia, Canada is the developer of this technology. Kobelco-Eco Solutions Co., Ltd. based in Japan is the exclusive licensee for this technology. Currently, this technology is only available in Japan. Powertech Labs Inc. indicated that that there are no plans at this time to license the technology in North America. No information regarding process details, performance data, or costs could be obtained directly from the technology vendor. Vendor contact information is provided in Section 5.0.

### 3.2.3    Subcritical Water Oxidation

Subcritical water oxidation is an ex situ process used to treat POPs. For this process, POPs must be in a liquid form and may require extraction into acetone prior to treatment. Within the treatment system, preheated water and sodium hydroxide solution, high-pressure oil and oxygen react in a reaction tower, at specific temperature and pressure (to 370°C and 26.7 megapascal). Carbon dioxide generated by the oxidation of the oil reacts with sodium hydroxide to produce sodium carbonate. When the specified conditions are reached inside the reaction tower, the contaminated waste stream to be treated replaces the oil, and decomposition of the contaminants occurs. Processed liquid that has completed the decomposition process is cooled and after depressurization, the liquid and gas are separated. The treated liquid is tested to confirm decomposition, while the gas is passed through an activated carbon unit prior to discharge.

> **TECHNOLOGY TYPE:**
> **THERMAL-CHEMICAL**
> **DEGRADATION**
>
> **POPs: ALDRIN, DIELDRIN, CHLORDANE, PCBs, DIOXINS AND FURANS**
>
> **MEDIUM: SOLID WASTE**
> **PRETREATMENT: EXTRACTION INTO A SOLVENT**
>
> *PILOT SCALE*
> *EX SITU*

A pilot plant in Japan has processed PCB contaminated waste streams for more than 3,500 hours without encountering difficulties. In 2005, a full-scale system was being constructed in Japan with a capacity of 2 tons per day. However, the current status of this system is not documented in the information identified and reviewed for this report (Ref. 44).

> **THE FACT SHEET PREPARED BY IHPA IS AVAILABLE AT**
> HTTP://WWW.IHPA.INFO/RESOURCES/LIBRARY/

Subcritical water oxidation is a thermochemical technology that uses thermal degradation and sodium hydroxide to treat POPs. A pre-treatment stage is needed for treating contaminated soil to extract the contaminants in solution. This technology has been used at a pilot-scale to treat PCBs. Due to the high temperature requirement of this technology, other POPs could also be potentially treated using Subcritical water oxidation. The most recent application of this technology was in 2005 to treat a PCB-contaminated waste stream in Japan. The information for this technology was provided by John Vijgen (IHPA). No performance data, process details, or costs for this technology could be obtained directly from the technology vendor. Currently, this technology is available only through the technology vendor Mitsubishi Heavy Industries of Japan. Vendor contact information is provided in Section 5.0.

## 3.3    Bench-Scale Technologies for Treatment of POPs

This section describes the technologies that have been implemented to treat POPs at the bench scale. Each subsection focuses on a single technology and includes a description of the technology and information about its application at specific sites. Fact sheets developed by IHPA contain additional details on some of these technologies and their applications. Links to the IHPA fact sheets are included in the appropriate subsections of this report.

### 3.3.1    Self-Propagating High-Temperature Dehalogenation

Self-propagating high-temperature dehalogenation (SPHTD) is an ex situ technology that has been tested to treat soil containing HCB contamination.

> **THE FACT SHEET PREPARED BY IHPA IS AVAILABLE AT**
> HTTP://WWW.IHPA.INFO/RESOURCES/LIBRARY/.

For SPHTD to operate, HCB containing soil would be mixed with calcium hydride or calcium metal, and the mixture is placed in a reaction chamber containing a tungsten coil. Addition of purified argon gas causes the reaction chamber to become pressurized, and an electrical pulse to the tungsten coil initiates the reaction. Once initiated, the reductive reactions that occur in the reaction chamber are exothermic and self-propagating. The reaction mixture can reach a temperature of 1,400°C, which creates thermochemical conditions that convert HCB to calcium chloride, carbon, and hydrogen (Ref. 71).

> **TECHNOLOGY TYPE: THERMAL-CHEMICAL DEGRADATION**
>
> **POPs TREATED: HCB**
>
> MEDIUM: POP STOCKPILES
>
> *BENCH SCALE*
> *EX SITU*

SPHTD has been tested at bench scale using materials contaminated with HCB, but no bench-scale test results were available in the information identified and used for this report (Ref. 41). The information sources identified and used to prepare this report also did not provide information about application of the technology at the pilot or full scale.

Self-propagating high-temperature dehalogenation is a thermal-chemical degradation technology that has been used to treat HCB-contaminated soil at bench scale. The SPHTD technology is being developed by Centro Studi Sulle Reazioni Autopropaganti, University of Cagliari in Italy and is not currently commercially available. Further technology information can be obtained by contacting the technology developer using the information provided in Section 5.0.

### 3.3.2    TDT-3R™

TDT-3R™ is an ex situ technology that has been tested for treatment of high- and low-strength soils containing HCB contamination.

> **TECHNOLOGY TYPE: THERMAL DEGRADATION**
>
> **POPs TREATED: HCB**
>
> **PRETREATMENT: THERMAL DESORPTION**
>
> MEDIUM: SOIL
>
> *BENCH SCALE*
> *EX SITU*

The TDT-3R™ technology uses a continuous low-temperature thermal desorption process conducted in the absence of air. The main component of this process is a specially designed, indirectly fired, horizontally arranged rotary kiln. Contaminated soil is heated in the kiln to a temperature typically between 300 and 350°C under an applied vacuum of 0 to 50 Pascal (Pa). In some instances, the kiln is heated to higher temperatures when POPs are being treated. The contaminants in the soil desorb and vaporize in the kiln. The vaporized contaminants are recovered from the kiln and combusted in a thermal oxidizer for at least 2 seconds at a temperature exceeding 1,250°C. Off-gas from the thermal oxidizer is rapidly cooled, passed through a wet gas multi-venturi scrubber, and discharged. Process water from the scrubber is treated and discharged. Treated soil exiting the kiln is cooled indirectly and removed (Refs. 44 and 64).

TDT-3R™ has been implemented at a bench scale in Gare, Hungary, to treat 100 kilograms (kg) of soil contaminated with HCB. Treatment occurred at a temperature of 450°C under a vacuum of 30 Pa. The technology reduced the soil's HCB concentration from 1,215 to 0.1 mg/kg (Ref. 65).

> **THE FACT SHEET PREPARED BY IHPA IS AVAILABLE AT** HTTP://WWW.IHPA.INFO/RESOURCES/LIBRARY/.

TDT-3R™ is a thermal degradation technology that uses continuous low-temperature desorption and a thermal oxidizer to treat contaminated soil. The technology has been used at bench scale to treat HCB contaminated soil. Due to the high temperature (temperature exceeding 1,250°C) requirement of this technology, other POPs could also be potentially treated using TDT-3R™. TDT-3R™ is marketed by Thermal Desorption Technology Group LLC in the US and its European subsidiaries. This firm has developed pilot-scale kilns that operate with throughput of 0.1 tons per hour. A larger thermal desorption technology kiln that would operate with a throughput of 4 m³/hour has been engineered and designed. According to the vendor, a conceptual design has been developed for a kiln with a throughput of 70 tons per hour (Ref. 64). Further technology information can be obtained by contacting the vendor using the information provided in Section 5.0.

## 4.0    INFORMATION SOURCES

The following web-based information sources were used during the preparation of this report. Additional information on POPs can be obtained from the websites identified below as well as from the references listed in Section 6.0.

**Stockholm Convention**
http://www.pops.int/

**International HCH and Pesticides Association**
http://www.ihpa.info/resources/library/

**Science and Technology Advisory Panel of the Global Environmental Facility**
http://stapgef.unep.org/

**U.S. Environmental Protection Agency**
http://www.clu-in.org/POPs
http://www.clu-in.org/acwaatap
http://www.epa.gov/oppfod01/international/pops.htm

**United Nations**
http://www.basel.int
http://www.chem.unep.ch/pops/
http://www.gpa.unep.org/pollute/organic.htm
http://www.who.int/iomc/groups/pop/en/
http://www.unido.org/doc/29487
http://www.unece.org/env/lrtap/welcome.html

**Other Sources**
http://www.africastockpiles.org/
http://www.fao.org/ag/AGP/AGPP/Pesticid/Disposal/index_en.htm
http://www.sdpi.org/research_Programme/environment/Hazardous_Waste_Management.htm#2
http://www.ipen.org

## 5.0    VENDOR CONTACTS

*Full-Scale Technologies for Treatment of POPs*

**Anaerobic Bioremediation Using Blood Meal for Treatment of Toxaphene in Soil and Sediment**
Mr. Harry L. Allen III, Ph.D.
US EPA Environmental Response Team
MS-101, Building 18
2890 Woodbridge Avenue
Edison, NJ 08837
Telephone:  (732) 321-6747
Fax: (732) 321-6724
Email: allen.harry@epa.gov

**Base Catalyzed Decomposition**
Mr. Terrence Lyons
US EPA National Risk Management Research Laboratory
26 West Martin Luther King Drive
Cincinnati, OH 45268
Telephone:  (513) 569-7589
Fax: (513) 569-7676

Mr. Charles Rogers
BCD Group Inc.
Cincinnati, OH
Telephone: (513) 385-4459

**DARAMEND®**
Dr. Alan G. Seech or Mr. David Raymond
Adventus Remediation Technologies, Inc.
1345 Fewster Drive
Mississauga, Ontario, Canada L4W 2A5
Telephone:  (905) 273-5374, Extension 221
Mobile: (416) 917-0099
Fax: (905) 273-4367
Email: info@AdventusGroup.com
Website: http://www.adventusgroup.com/

**Gas-Phase Chemical Reduction (GPCR™)**
Bennett Environmental Inc.
1540 Cornwall Road, Suite 208
Oakville, Ontario, Canada L6J 7W5
Telephone: (905) 339-1540
Fax: (905) 339-0016
Email: info@bennettenv.com
Website: http://www.bennettenv.com

**Gene Expression Factor (bioremediation)**
Christopher Young
GeoSolve, Inc.
137 Cross Center Road, #143
Denver, North Carolina 28037
Telephone: 704-489-6538
Email: cyoung2281@aol.com

**GeoMelt™**
Mr. Bret Campbell or Mr. Keith Witwer
IMPACT Services, Inc.
GeoMelt Division
1135 Jadwin Avenue
Richland, WA  99352
Telephone:  (509) 942-1114
Fax: (509) 99942-1122
Email:  Bret.Campbell@impactserviceinc.com
or Keith.Witwer@impactserviceinc.com

**In Situ Thermal Desorption (ISTD)**
Mr. Ralph Baker
TerraTherm, Inc.
356 Broad Street
Fitchburg, MA  01420
Telephone:  (978) 343-0300
Fax:  (978) 343-2727
Email:  rbaker@terratherm.com
Website:  www.terratherm.com

**Mechanochemical Dehalogenation (MCD™)**
Mr. Bryan Black
Environmental Decontamination Ltd.
P.O. Box 58-609
Greenmount
Aukland, New Zealand
Telephone:  (649) 274-9862
Fax:  (649) 274-7393
Email:  bryan@manco.co.nz
Website:  http://edl.net.nz

Mr. Volker Birke
Tribochem
Georgstrasse 14, D-31515 Wunstdrof, Germany
Telephone:  49 5031 6 73 93
Fax:  49 5031 88 07
Email:  birke@tribochem.com
Website:  www.tribochem.com

**SRL Plasma Pty Ltd**
PO Box 119 Narangba
Qld. 4504 Australia
Telephone: 61 7 3203 3400
Fax: +61 7 3203 3450
Email: nevillet@srlplasma.com.au
Website: http://www.plascon.com.au/

**Radicalplanet® Technology**
Head Office: 1-21-8 Sakae Naka-ku,
Nagoya City, Aichi 460-0008
Telephone: +81-52-222-8333
Fax: +81-52-702-6620
info@radicalplanet.co.jp

**Solvated Electron Technology**
Commodore Advanced Sciences, Inc
Jonathan Rogers
507 Knight Street
Suite B
Richland, WA 99352
Telephone: 865.483.9619
Fax: 509.943.2910
Email: jonrogers@commodore.com
Website: http://www.commodore.com

**Sonic Technology**
Mr. Claudio Arato
Sonic Environmental Solutions Inc.
1066 West Hastings Street, Suite 2100
Vancouver, British Columbia, Canada
V6E 3X2
Telephone: (604) 736-2552
Fax: (604) 736-2558
Email: carato@sonictsi.com
Website: www.sonicenvironmental.com

*Pilot-Scale Technologies for Treatment of POPs*

**Reductive Heating and Sodium Dispersion**
Keith Lee
Powertech Labs
12388 88th Ave, Surrey,
BC V3W 7R7, Canada
Telephone: 604-590-7438
Email: keith.lee@powertechlabs.com

*Bench-Scale Technologies for Treatment of POPs*

**Self-Propagating High-Temperature Dehalogenation**
Dr. Ing. Giacomo Cao Centro Studi Sulle
Reazioni Autopropaganti Dipartimento di
Ingegneria Chimica e Materiali Piazza d'armi
09123 Cagilari Italy
Telephone: 39-070-6755058
Fax: 39-070-6755057
Email: cao@crs4.it

**TDT-3R™**
Mr. Edward Someus
Terra Humana Clean Technology Engineering Ltd.
1222 Budapest, Szechenyi 59
Hungary
Telephone: (36-20) 201 7557
Fax: (36-1) 424 0224
Email: edward@terrenum.net
Website: http://www.terrenum.net

*Technologies Identified in 2005 Report to Treat POPs but Not Commercially Available*

**Xenorem™**
Mr. Brad Yops
Technology Transfer Corporation
University of Delaware
Newark, DE 19716
Telephone: (302) 831-0147
Website: http://www.udel.edu/

**Supercritical Water Oxidation**
Dave Ordway
Telephone: 858-455-3568
Email: david.ordway@gat.com
Email: info@turbosynthesis.com

Special acknowledgment is given to members of the International HCH and Pesticides Association (IHPA), and other remediation professionals for their cooperation, thoughtful suggestions, and support during the preparation of this report.  Contributors to the report include the following individuals:

Alan G. Seech, Adventus Remediation Technologies, Inc.
Brad Yops, Technology Transfer Corporation, University of Delaware
Bryan Black, Environmental Decontamination Ltd.
Bret Campbell, IMPACT Services, Inc.
Carl V. Mackey, Washington Group International
Charles Rogers, BCD Group, Inc.
Christine Parent, California Department of Toxic Substances Control-
Christopher Young, BioTech Restorations
Claudio Arato, Sonic Environmental Solutions, Inc.
Edward Someus, Terra Humana Clean Technology Engineering Ltd.
Giacomo Cao, Centro Studi Sulle Reazioni Autopropaganti
John Fairweather, Thermal and Chemical Soil Remediation Ltd
John Vijgen, IHPA
Kevin Finucane, AMEC Earth and Environmental, Inc.
Jonathan Rogers, Commodore Advanced Sciences, Inc
Ralph Baker, TerraTherm, Inc.
Tedd E. Yargeau, California Department of Toxic Substances Control
Volker Birke, Tribochem

## 6.0    REFERENCES

1.    Adventus Americas, Inc. 2009. "Representative DARAMEND® Project Summaries." Online Address: http://www.adventusgroup.com/projects/proj_daramend.shtml.

2.    Adventus Americas, Inc. April 2009. "Bioremediation of Pesticides." Online Address: http://www.adventusgroup.com/pdfs/tech_bullet/DARAMEND%20Bioremediatoin%20of%20PESTICIDES.pdf.

3.    Allen, H.L., and others. 2002. "Anaerobic Bioremediation of Toxaphene Contaminated Soil – A Practical Solution." 17th World Congress of Soil Science, Symposium No. 42, Paper No. 1509. Thailand. August 14 through 21.

4.    Allen, H.L., US EPA Environmental Response Team. January 25, 2005. Email to Younus Burhan, Tetra Tech EM Inc., Regarding US EPA Comments on Draft Blood Meal Fact Sheet.

5.    American Chemical Society. May 2000. "Herbicide and Pesticide Destruction." Symposium on Emerging Technologies: Waste Management in the 21st Century. San Francisco, California.

6.    Assembled Chemical Weapons Alternative (ACWA). 2009. "About ACWA." Online Address: http://www.pmacwa.army.mil/about/index.html.

7.    Agency for Toxic Substances and Disease Registry (ATSDR). 2009. Website for ATSDR. Online Address: http://www.atsdr.cdc.gov.

8.    ATSDR. March 2002. "Public Health Reviews of Hazardous Waste Thermal Treatment Technologies." Online Address: http://www.atsdr.cdc.gov/hac/thermal/thermal1.html.

9.    Baker, R., TerraTherm, Inc. October 27 and November 8, 15, 24, and 29, 2004. Emails to Chitranjan Christian, Tetra Tech EM Inc.

10.   Baker, R., and M. Kuhlman. 2002. "A Description of the Mechanisms of In Situ Thermal Destruction (ISTD) Reactions." Second International Conference on Oxidation and Reduction Technologies for Soil and Groundwater (ORTs-2). Toronto, Ontario, Canada. November 17 through 21.

11.   Baker, R.S., D. Tarmasiewicz, J.M. Bierschenk, J. King, T. Landler and D. Sheppard. 2007. Completion of In Situ Thermal Remediation of PAHs, PCP and Dioxins at a Former Wood Treatment Facility. *2007 International Conference on Incineration and Thermal Treatment Technologies (IT3)*, May 14-18, 2007, Phoenix, AZ. Air & Waste Management Association, Pittsburgh, PA.

12.   The Basel Convention on the Control of Transboundary Movements of Hazardous Wastes and their Disposal. 2010. Website for the Basel Convention. Online Address: http://www.basel.int.

13.   Birke, V. 2002. "Reductive Dehalogenation of Recalcitrant Polyhalogenated Pollutants Using Ball Milling." Proceedings of the Third International Conference on Remediation of Chlorinated and Recalcitrant Compounds. Monterey, California.

14.   Boethling, R.S. and D. Mackay. 2000. "Handbook of Property Estimation Methods for Chemicals." Environmental and Health Sciences. Boca Raton, Florida. Lewis Publishers

15.   CerOx™ Corporation. 2005. Process Technology Overview. Online address is not available.

16. Donnelly, P.K., Hedge, R.S., and Fletcher, J.S. 1994. "Growth of PCB-Degrading Bacteria on Compounds from Photosynthetic Plants." Chemosphere. Volume 28, Number 5. Pages 981-988.

17. The Federal Remediation Technologies Roundtable (FRTR). 2009. "Remediation Technologies Screening Matrix and Reference Guide, Version 4." Online Address: http://www.frtr.gov/matrix2/top_page.html.

18. U.S. Department of Health and Human Services (DHHS). 2005. "Toxicological Profile Information Sheet." Agency for Toxic Substances and Disease Registry (ATSDR). Online Address: http://www.atsdr.cdc.gov/toxpro2.html.

19. US EPA. September 1994. "Eco Logic International Gas-Phase Chemical Reduction Process-The Thermal Desorption Unit." Superfund Innovative Technology Evaluation Program. EPA/540/AR-94/504. Online Address: http://www.epa.gov/nrmrl/lrpcd/site/reports/540ar93522/540ar93522.pdf.

20. U.S. Environmental Protection Agency (US EPA). 1996. "Cost and Performance Summary Report, Bioremediation at the Stauffer Management Company Superfund Site, Tampa, Florida." Office of Solid Waste and Emergency Response.

21. US EPA. 1996. Site Technology Capsule: "GRACE Bioremediation Technologies DARAMEND® Bioremediation Technology." Superfund Innovative Technology Evaluation Program. EPA/540/R-95/536.

22. US EPA. September 2000. "Cost and Performance Summary Report, Bioremediation at the Stauffer Management Company Superfund Site, Tampa, Florida." Office of Superfund Remediation and Technology Innovation.

23. US EPA. August 2000. "Potential Applicability of Assembled Chemical Weapons Assessment Technologies to RCRA Waste Streams and Contaminated Media." Office of Solid Waste and Emergency Response, Technology Innovation Office. EPA-R-00-004. Online Address: http://www.epa.gov/tio/download/remed/acwatechreport.pdf.

24. US EPA. April 2002. "Persistent Organic Pollutants. A Global Issue. A Global Response." Office of International Affairs. EPA160-F-02-001. Online Address: http://www.epa.gov/oia/toxics/pop.pdf.

25. US EPA. July 2004. "Field Evaluation of TerraTherm In Situ Thermal Destruction (ISTD) Treatment of Hexachlorocyclopentadiene." Office of Research and Development, Superfund Innovative Technology Evaluation Program. EPA/540/R-05/007. Online Address: http://www.epa.gov/ORD/NRMRL/pubs/540r05007/540R05007.pdf.

26. US EPA. March 2004. "In Situ Thermal Treatment of Chlorinated Solvents: Fundamentals and Field Applications." EPA 542-R-04-010. Online Address: http://www.epa.gov/tio/download/remed/epa542r04010.pdf.

27. US EPA. 2005. Website on Persistent Organic Pollutants (POP). Office of Pesticide Programs. Information Downloaded on January 5. Online Address: http://www.epa.gov/international/toxics/pop.htm.

28. US EPA. January 2007. "Combined Mechanical/Chemical Process Removes POPs from Soil and Sediment." Technology News and Trends. Issue 28. EPA 542-N-06-007. Online Address: http://www.clu-in.org/products/newsltrs/tnandt/view.cfm?issue=0107.cfm.

29. US EPA. January 2007. "EPA Evaluates Cost and Performance of Blood Meal-Enhanced Anaerobic Bioremediation of Toxaphene-Contaminated Soil." Technology News and Trends. EPA 542-N-06-007. Issue 28. Online Address: http://www.clu-in.org/products/newsltrs/tnandt/view.cfm?issue=0107.cfm.

30. Finucane, K., Geomelt. May 23, 2005. Emails to Ellen Rubin, US EPA, Office of Superfund Remediation Technology Innovation, Regarding Geomelt ICV Process Description and Photo, and Results for WCS Rock Flats and WCS Commercial TSCA Cleanup.

31. General Atomics. 2005. "Advanced Process Systems Division." Online Address: http://www.ga.com/atg/aps.

32. GeoSolve, Inc. 2009. "Factor Description." Online Address: http://www.geosolve-inc.com/biotech.html.

33. Getman, G.D. 2001. "Closure Report for SoLV Process Treatment of Soils Excavated from the Pennsylvania Air National Guard Site At Harrisburg International Airport Harrisburg, Pennsylvania."

34. Gilbert, E.S. and D.E. Crowley. 1997. "Plant Compounds that Induce Polychlorinated Biphenyl Biodegradation by Anthrobacter sp. Strain B1B." Applied and Environmental Microbiology. Volume 63, Number 5. Pages 1933-1938.

35. Global Security. Weapons of Mass Destruction. Army Facilities. 2005. "Newport Chemical Depot (NECD), Newport, Indiana." Online Address: http://www.globalsecurity.org/wmd/facility/newport.htm.

36. Gray, N.C.C., AstraZeneca Group PLC. December 15, 2004. Telephone Conversation with Younus Burhan, Tetra Tech EM Inc.

37. Gray, N.C.C., P.R. Cline, A.L. Gray, B. Byod, G.P. Moser, H.A. Guiler, and D.J. Gannon. 2002. "Bioremediation of a Pesticide Formulation Plant." Proceedings of the Third International Conference on Remediation of Chlorinated and Recalcitrant Compounds. Monterey, California.

38. Haley & Aldrich, 1997. Demonstration Test Report, Thermal Wells, In Situ Thermal Desorption Technology, Missouri Electric Works Site, Cape Girardeau, Missouri, Nov. 1997, Rochester, New York. p. 31 and Appendix B, p. 2.

39. Hurt, K. 2005. "Successful Full Scale Phytoremediation of PCB and TPH Contaminated Soil," The Third International Phytotechnologies Conference, Atlanta, Georgia. April 19 - 22.

40. Impact Services, Inc. 2005. "GeoMelt® Technology Description." Online Address: http://www.impactservicesinc.com/divisions/geomelt.html.

41. Ing, G.C., Centro Studi sulle Reazioni Autopropaganti. December 12, 2004. Email to Younus Burhan, Tetra Tech EM Inc.

42. International Centre for Science and High Technology. 2007. Website for International Centre for Science and High Technology; United Nations Industrial Development Organization. Online Address: http://www.ics.trieste.it.

43. International HCH and Pesticides Association (IHPA). June 2005. "Gas-Phase Chemical Reduction." Online Address: http://www.ihpa.info/docs/library/reports/Pops/FactSheet%20SBC_GPCR%20_251105_DEF.pdf.

44. IHPA. 2002. "NATO/CCMS Pilot Study Fellowship Report." Evaluation of Demonstrated and Emerging Remedial Action Technologies for the Treatment of Contaminated Land and Groundwater (Phase III). Online Address: http://www.ihpa.info/docs/library/libraryNATO.php.

45. Ishii, Y., Kawai, T., Nobuaki, H., Takehara, H., Yara, H., and Y. Tokashiki. March 2007. "POPs Contaminated Soil Treatment with 'Reductive Heating and Sodium Dispersion Method' and Its Recycling for Material of Green Planting." Journal of Environmental Science for Sustainable Society. Volume 1. Pages 11 through 14.

46. Johnson, L., General Atomics. February 15, 2005. Telephone Conversation with Chitranjan Christian, Tetra Tech EM Inc.

47. Leigh, M., Fletcher, J., Nagle, D.P., Prouzova P., Mackova, M. and Macek, T. 2003. "Rhizoremediation of PCBs: Mechanistic and Field Investigations." Proceedings of the International Applied Phytotechnologies Conference. Chicago, Illinois.

48. Lyons, T., US EPA National Risk Management Research Laboratory. January 19 and August 10, 2005. Email to Younus Burhan, Tetra Tech EM Inc.

49. March, J. 1985. "Advanced Organic Chemistry. Reactions, Mechanisms, and Structure." Third Edition. New York. John Wiley & Sons.

50. Moklyachuk, L., Sorochinky, B. and Kulakow, P.A. 2005. "Phytotechnologies for Management of Radionucleide and Obsolete Pesticide Contaminated Soil in Ukraine," The Third International Phytotechnologies Conference, Atlanta, Georgia. April 19 - 22.

51. Noma, Y. 2009. Japanese Pesticides Treatment Fact Sheets, National Institute for Environmental Studies, Research Center for Material Cycles and Waste Management, Ibaraki, Japan. http://www.ihpa.info/docs/library/other/jp_factsheet/Fact_sheet%28Radicalplanet_Technology%29 1.pdf.

52. Nurzhanova, A., Kulakow, P., Rubin, E., Rakhimbaev, I., Sedlovsky, A., Zhambakin, K., Kalygin, S., Kalmykov, E. L. and Erickson. L. 2005. "Monitoring Plant Species Growth in Pesticide Contaminated Soil," The Third International Phytotechnologies Conference, Atlanta, Georgia. April 19 - 22.

53. Pan, D., Bennett Environmental Inc. January 18, 2005. Telephone Conversation with Younus Burhan, Tetra Tech EM Inc.

54. Phillips, T., G. Bell, D. Raymond, K. Shaw, and A. Seech. 2001. "DARAMEND® Technology for In Situ Bioremediation of Soil Containing Organochlorine Pesticides." 6th International HCH and Pesticides Forum, Poznan, Poland. March 20 - 22.

55. Raymond, D., Adventus Remediation Technologies, Inc. August 25, 2004. Telephone Conversation with Younus Burhan, Tetra Tech EM Inc.

56. Rogers, C., BCD Group Inc. December 9 and 13, 2005. Email correspondence with Younus Burhan, Tetra Tech EM Inc.

57. Shaw Environmental Inc. and Environmental Decontamination Ltd. May 15, 2008. "MechanoChemical Destruction Supplemental Treatability Study, PCB Hot Spot – Soil Stockpiles 12 and 15, Hunters Point Shipyard, San Francisco, California." Final Report.

58. Shimme, K., Radicalplanet Research Institute Co. Ltd. October 8, 2009. E-mail correspondence with Younus Burhan, Tetra Tech EM Inc.

59. Sonic Environmental Solutions Inc. 2005. Website for Sonic Technology. Online Address: http://www.sesi.ca.

60. Stegemeier, G.L., and Vinegar, H.J. 2001. "Thermal Conduction Heating for In Situ Thermal Desorption of Soils," Chapter 4.6, pages 1-37. Chang H. Oh (ed.), Hazardous and Radioactive Waste Treatment Technologies Handbook, CRC Press, Boca Raton, FL.

61. Stockholm Convention on Persistent Organic Pollutants (POPs). 2009. Website for the Stockholm Convention. Online Address: http://www.pops.int.

62. TerraTherm Environmental Services. November 1999. "Naval Facility Centerville Beach, Technology Demonstration Report, In Situ Thermal Desorption."

63.  TerraTherm, Inc. 2005. "Process Description In Situ Thermal Desorption (ISTD)." Online Address: http://www.terratherm.com.

64.  Thermal Desorption Technology Group LLC of North America. 2005. Website for Terra Humana Clean Technology Engineering Ltd. Online Address: http://www.terrenum.net.

65.  Thermal Desorption Technology Group, Terra Humana Clean Technology Engineering Ltd. December 10, 2004. "Summary Report of the TDT-3R Treatment - Latest Five Years – Projects 2000-2004."

66.  Thiess Services NSW. June 2004. "Proof of Performance Report, FCC Remediation, Mapua, New Zealand."

67.  Turbosystems Engineering Inc. 2005. Web Page for Supercritical Water Oxidation Technology. Online Address: http://www.turbosynthesis.com/summitresearch/sumhome.htm.

68.  United Nations Economic Commission for Europe (UNECE). 2009. Website for the United Nations Economic Commission for Europe. Online Address: http://www.unece.org/env/lrtap.

69.  United Nations Environment Programme (UNEP). 2005. Website for the Secretariat of the Basel Convention. Online Address: http://www.basel.int.

70.  UNEP. May 2005. "Technical Guidelines for the Environmentally Sound Management of Persistent Organic Pollutant Wastes." UNEP/POPS/COP.1/11. Online Address: http://www.pops.int/documents/meetings/cop_1/meetingdocs/en/cop1_11/COP_1_11.pdf.

71.  UNEP. January 2004. STAP of the GEF. "Review of Emerging, Innovative Technologies for the Destruction and Decontamination of POPs and the Identification of Promising Technologies for Use in Developing Countries." GF/8000-02-02-2205. Online Address: http://www.basel.int/techmatters/review_pop_feb04.pdf.

72.  UNEP. October 2003. Science and Technology Advisory Panel (STAP) of the Global Environmental Facility. "Report of the STAP/GEF POPs Workshop on Non-Combustion Technologies for the Destruction of POPs Stockpiles." Online Address: http://www.basel.int/techmatters/review_pop_feb04.pdf.

73.  White, J. C., Mattina, M. I., Eitzer, B. D., Isleyen, M., Parrish, Z. D. and Gent, M. P.N. 2005. "Enhancing the Uptake of Weathered Persistent Organic Pollutants by Cucurbita pepo," The Third International Phytotechnologies Conference, Atlanta, Georgia. April 19 - 22.

74.  Whitfield, M., Rutter, A., Reimer, K. J., and Zeeb, B., 2008. "The Effects of Repeated Planting, Planting Density, and Specific Transfer pathways on PCB uptake by Cucurbita pepo grown in field conditions," Science of the Total Environment, September 10.

75.  Zeeb, B., Whitfield, M. and Reimer, K. J. 2005. "In situ Phytoextraction of PCBs from Soil: Field Study," The Third International Phytotechnologies Conference, Atlanta, Georgia. April 19 - 22.

# APPENDIX A

**Chemical Structures, Uses and Effects of POPs listed under the Stockholm Convention and LRTAP**

# PESTICIDES

### Aldrin & Dieldrin

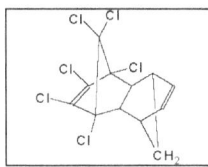

Aldrin and Dieldrin are chemical compounds that were once used as insecticides. Aldrin and Dieldrin are persistent in the environment and resistant to biodegradation and abiotic transformation. They both undergo bioaccumulation and bio-magnify in terrestrial and aquatic ecosystems. Aldrin and Dieldrin are toxic to humans and cause damage to the liver and immune system. They are also carcinogenic to certain animals (Ref. 7).

### Alpha-hexachlorocyclohexane & Beta-hexachlorocyclohexane

Alpha and beta-hexachlorocyclohexane (HCH) are by-products from the production of Lindane. Alpha and beta-HCH are both persistent in the environment; however, beta-HCH is found to be more persistent than alpha-HCH. Both HCHs bioaccumulate in aquatic and terrestrial ecosystems and can transport over long ranges in the atmosphere. Alpha-HCH and beta-HCH are considered to be human carcinogens that affect the human reproductive, neurological and immune systems (Ref. 7).

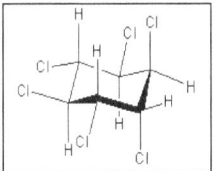

### Chlordane

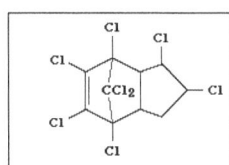

Chlordane is an organochlorine compound that was once used as a pesticide. It is extremely persistent in the environment and has the potential to bioaccumulate in aquatic ecosystems. There is not sufficient evidence to consider chlordane as a human carcinogen; however, chlordane is toxic to humans and affects the liver, digestive and nervous systems (Ref. 7).

### Chlordecone

Chlordecone is a manufactured chemical that is present in pesticides and insecticides. Chlordecone is persistent in the environment, has a tendency to bioaccumulative and bio-magnify in terrestrial and aquatic food chains, is readily absorbed in soil and sediments, and has a high resistance to biodegradation. Chlordecone is highly toxic to humans, as well as aquatic organisms, and damages the musculoskeletal, liver, neurological and immune systems. Chlordecone is considered as a human carcinogen (Ref. 7).

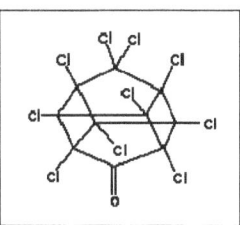

### Dichlorodiphenyltrichloroethane (DDT)

Dichlorodiphenyltrichloroethane (DDT) was a widely used pesticide and insecticide that controlled agricultural crop pests as well as insects that carried human diseases. DDT is persistent in the environment, which contributes to the bioaccumulation and bio-magnification effect that DDT has on organisms in the environment. DDT also has the potential to undergo long-range global transport through the air. DDT is listed as a "probable" carcinogen to humans and can cause damage to the lungs and respiratory system if inhaled (Ref. 7).

### Endosulfan

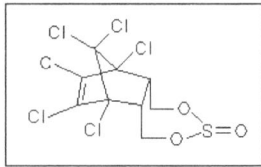

Endosulfan is a synthetic organochlorine compound that is widely used in agricultural insecticides. It is persistent in the environment, bioaccumulative, and has the potential for long-range environmental transport. Endosulfan is highly toxic to all organisms and humans; it affects the neurological, reproductive and developmental systems in humans (Ref. 7). Based on recent data and evaluations, as of June 2010, EPA is taking action to eliminate all uses of endosulfan in the US; additional information is available at
http://www.epa.gov/pesticides/reregistration/endosulfan/endosulfan-cancl-fs.html.

### Endrin

Endrin was primarily used as an insecticide and a rodenticide. Endrin is extremely persistent in the environment because it adsorbs strongly to soil particles and is practically immobile, which contributes to endrin's high bioaccumulation capabilities. Unlike other organochlorines, endrin has a relatively low bio-magnification factor. Endrin affects the nervous system in humans, but is not considered to be a human carcinogen (Ref. 7).

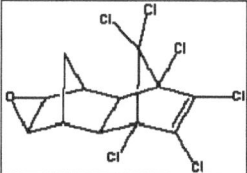

### Heptachlor

Heptachlor is an organochlorine that has been used as an insecticide. Heptachlor is persistent in the environment and adsorbs strongly to soil sediments, which contributes to a high bioaccumulation factor. Heptachlor has a relatively low bio-magnification factor. It is listed as a "possible" human carcinogen that affects the liver, nervous and reproductive systems (Ref. 7).

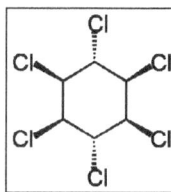

### Hexachlorobenzene (HCB)

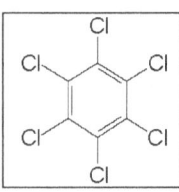

Hexachlorobenzene (HCB) is classified as a chlorinated hydrocarbon and was used as a fungicide for agricultural seed treatment. HCB is one of the most persistent chemicals found in the environment due to its chemical stability and high resistance to degradation. HCB significantly bioaccumulates in both terrestrial and aquatic food chains. For humans, HCB is considered a "probable" human carcinogen and is also an animal carcinogen (Ref. 7).

### Lindane

Lindane is an organochlorine that was used in agricultural insecticides and in the pharmaceutical treatment of lice or scabies. Lindane is persistent in the environment and has the potential to bioaccumulate and bio-magnify in terrestrial and aquatic ecosystems. There is significant evidence for lindane to be considered as carcinogenic but is currently listed as a "possible" human carcinogen (Ref. 7).

### Mirex

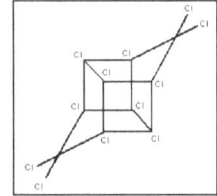

Mirex is a chlorinated hydrocarbon that was used as an insecticide and is also found in flame retardants. Mirex, like HCB, is very persistent in the environment due to its high resistance to chemical and biological degradation. This contributes to high bioaccumulation and bio-magnification factors in terrestrial and aquatic ecosystems. Mirex is considered to be an animal carcinogen and can affect the liver in humans (Ref. 7).

## Toxaphene

Toxaphene, a complex mixture of hundreds of organic compounds, was used as an insecticide. It is persistent in the environment and has the potential to bioaccumulate and bio-magnify in terrestrial and aquatic ecosystems. Toxaphene can be transported over very long distances in the atmosphere. Toxaphene is a human carcinogen, which affects the kidneys, lungs and nervous system (Ref. 7).

# INDUSTRIAL CHEMICALS OR BY-PRODUCTS

## Polychlorinated biphenyls (PCBs)

Polychlorinated biphenyls (PCBs) were one of the most widely manufactured industrial chemicals in the US. PCBs are extremely persistent in the environment, bioaccumulate significantly in terrestrial and aquatic ecosystems, and have a very high resistance to environmental degradation. In humans and animals, PCBs are considered to be a carcinogen and can affect the liver and kidneys (Ref. 7).

## Dioxins & Furans

Dioxins and furans are by-products associated with the production of organochlorides. Dioxins and furans are persistent in the environment and are resistant to biodegradation. They also have a great potential to bioaccumulate in terrestrial and aquatic ecosystems. Several different forms of dioxins and furans are considered to be "possible" human carcinogens (Ref. 7).

## Octabromodiphenyl ether, Pentabromodiphenyl ether (penta-BDE), Hexabromobiphenyl & Hexabromocyclododecane (HBCDD)

Octabromodiphenyl ether, pentabromodiphenyl ether (penta-BDE), hexabromobiphenyl and hexabromocyclododecane (HBCDD) are all classified as brominated flame retardants (BFRs). BFRs have been used as industrial chemicals to produce foam for furniture and upholstery and casings for electronic goods. BFRs are toxic and persistent in the environment. Under the Stockholm Convention, the POPs review committee provided experimental evidence that the bioaccumulation of high blood level of BFRs in women of childbearing age could potentially harm the women and their unborn children.
.

## Pentachlorobenzene

Pentachlorobenzene is persistent, bioaccumulative, and toxic to humans and aquatic organisms. Pentachlorobenzene also has a high bio-magnification potential and can undergo long-range transport in the air.  In humans, pentachlorobenzene affects the central nervous system, liver, kidneys and reproductive system (Ref. 7).

## Perfluorooctane sulfonate (PFOS)

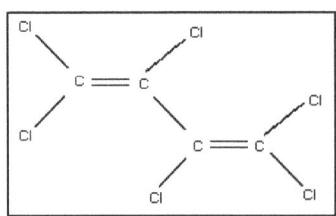

Perfluorooctane sulfonate (PFOS) is a widely used industrial chemical that is found in paints, polishes, leathers and fire retardants in the form of a fire fighting foam. PFOS is easily absorbed, bioaccumulative, persistent in the environment and toxic to humans and wildlife. PFOS also has the ability to transport over long distances in the environment (Ref. 7).

## Hexachlorobutadiene (HCBD)

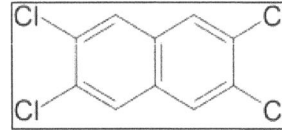

Hexachlorobutadiene (HCBD) is mainly used to make rubber compounds. It is also used as a solvent, and to make lubricants, in gyroscopes, as a heat transfer liquid, and as a hydraulic fluid.  HCBD is also a by-product of chemical processing. HCBD has the potential to transport over long distances through water, soil or the atmosphere. Limited data is available about the bioaccumulation and biomagnification abilities of HCBD but it is predicted to bioaccumulate only in aquatic ecosystems.  HCBD is considered a "possible" human carcinogen (Ref. 7).

## Polychlorinated naphthalenes (PCN)

Polychlorinated naphthalenes (PCNs) are chemical by-products that are created when chlorine reacts with naphthalene. This may occur during the production of coal tar. Limited data is available for PCNs, but available information indicates that PCNs likely have a low potential for bioaccumulation and biomagnification. PCNs are harmful to humans and affect the liver (Ref. 7).

## Short-chained chlorinated paraffins (SCCP)

Short-chained chlorinated paraffins (SCCPs) are industrial chemicals found in flame retardants, metal working fluids and in polyvinyl chlorinated (PVC) plastics. SCCPs are persistent, toxic (particularly to aquatic organisms), and undergo long-range transport in the environment.  In humans, SCCPs have the potential to harm a breast-fed child through bioaccumulation (Ref. 7).

# APPENDIX B

**Fact Sheet on
Anaerobic Bioremediation Using Blood Meal for the
Treatment of Toxaphene in Soil and Sediment**

## *Anaerobic Bioremediation Using Blood Meal for the Treatment of Toxaphene in Soil and Sediment*

**POPS-WASTES APPLICABILITY (REFS. 1 AND 5):**
Anaerobic Bioremediation Using Blood Meal was able to rapidly degrade toxaphene in soil to achieve cleanup goals in bench- and pilot-scale tests. Bench-scale tests have indicated that the technology is also effective in treating dichlorodiphenyltrichloroethane (DDT). Full-scale implementations have successfully treated several toxaphene-contaminated sites. The quantity of soil treated at these sites ranged from 250 to 8,000 cubic yards. This technology does not typically achieve greater than 90 percent contaminant reduction.

| | |
|---|---|
| **POPs Treated:** | Toxaphene and DDT |
| **Other Contaminants Treated:** | None |
| **Application:** | Ex-situ |

**TECHNOLOGY DESCRIPTION (REFS. 1 AND 5):**

**OVERVIEW**
This technology uses biostimulation to accelerate the degradation of toxaphene in soil or sediment. It involves the addition of biological amendments, including blood meal (nutrient) and phosphates (pH buffer), to stimulate native anaerobic microorganisms. Blood meal is a black powdery fertilizer made from animal blood. The typical dosage of blood meal and sodium phosphate is one percent by weight of contaminated soil. This is sometimes augmented with one percent by weight of starch to rapidly establish anaerobic conditions. The standard recipe uses monobasic and dibasic phosphate salts in equal proportions (monobasic:dibasic - 1:1) to maintain soil pH around 6.7. The low phosphate/starch recipe uses three times more dibasic than monobasic phosphates (monobasic:dibasic – 1:3) and maintains soil pH around 7.8.

The soil to be treated is mixed with amendments and water. Mixing methods including blending in a dump truck, mechanical mixing in a pit, and mixing in a pug mill have been used to produce homogeneous soil-amendment mixtures. The mixture is transferred to a cell with a plastic liner, and excess water is added to provide up to a foot of cover above the settled solids. The water provides a barrier that minimizes the transfer of atmospheric oxygen to microorganisms in the slurry, which helps maintain anaerobic conditions. The lined cell is covered with a plastic sheet to isolate the cell from the environment, and the slurry is incubated for several months. The slurry may be sampled periodically to measure treatment progress. Once treatment goals have been met, the cell is drained. The slurry is usually left in place, but it may be dried and used as fill material on site. The slurry also serves as a source of acclimated microorganisms for use at another toxaphene-contaminated site.

Anaerobic degradation of toxaphene usually results in the production of intermediates such as less chlorinated congeners of toxaphene. Further degradation of intermediates results in the production of carbon dioxide, methane, water, inorganic chlorides, and cell mass.

**STATUS AND AVAILABILITY (REFS. 2 AND 6):**
The technology has been implemented at full scale to treat toxaphene-contaminated sites. Four such sites are:

(1) The Laahty Family Dip Vat (LDV) site (253 cubic yards in one cell)
(2) The Henry O Dip Vat (HDV) site (660 cubic yards in two cells)
(3) The Gila River Indian Community (GRIC 1) site (3,500 cubic yards in four cells)
(4) The GRIC 2 site (8,000 cubic yards in five cells)

US EPA's Environmental Response Team (ERT) is the developer of the technology. The technology is unlicensed and is available through the ERT. The biological amendments (blood meal and monobasic and dibasic phosphates) are inexpensive and commercially available.

**Design (Refs 1, 5):**
Factors that need to be considered when designing an anaerobic bioremediation process using blood meal include:
- The presence of active toxaphene-degrading bacteria
- Soil characteristics
- Volume of soil to be treated
- Concentration of toxaphene in contaminated soil
- Cleanup goal
- Availability of space on site for the construction of treatment cells
- Odor mitigation requirements as determined by surrounding land use and the proximity of residences
- Need for agreements with landowners and community leaders
- Climate
- Security issues
- Availability of water

**THROUGHPUT (REFS. 1 AND 5):**
Throughput of a technology that does not operate like a batch processing plant is hard to define. Remediation involves a series of steps including construction, mix preparation, and treatment. Treatment is usually the slowest step. Factors that can influence treatment time include, the type of microbial communities present, amendment dosage, contaminant concentration, treatment goals, and the presence of inhibitors (such as very cold environments). In general, treatment time can vary from five weeks to two years.

**WASTES/RESIDUALS (REFS 2, 3 AND 6):**
Products of toxaphene degradation include lower-chlorinated chlorobornane congeners, chloride ions, cell mass, carbon dioxide, and methane. Chlorobornane congeners have been shown to degrade completely during treatment. However, treated soil can contain low concentrations (below cleanup goals) of unutilized toxaphene and lower-chlorinated chlorobornane congeners.

Gaseous wastes produced can include methane and hydrogen sulfide. Therefore, odor concerns should be considered. If treatment cells are not left in place at the end of remediation, solid wastes can include debris from the demolition of treatment cells and associated temporary facilities. Debris potentially contaminated with toxaphene will require testing to determine its hazardous nature in compliance with local, State, and Federal requirements prior to disposal.

**MAINTENANCE (REFS. 2 AND 6):**
- Periodic addition of water to treatment cells to maintain water level
- Maintaining treatment cells to prevent leaks
- Maintaining cover integrity
- Monitoring for gas buildup
- Monitoring for fugitive odors
- Soil sampling to monitor remedial progress

**LIMITATIONS (REFS. 2 AND 6):**
- The anaerobic process is affected by temperature. Spring and summer are the best periods for operation. This technology cannot be used in extremely cold climates.
- This technology requires a bench scale test to determine applicability at a given site, and to estimate treatment duration.
- At a minimum, five weeks are required for treatment.
- This technology typically does not achieve greater than 90 percent contaminant destruction.

- Blood meal accelerates the rate of reductive dechlorination of toxaphene, but does not affect the extent of dechlorination.
- Unfavorable soil chemistry can inhibit the process. Unfavorable soil chemistry may result from the presence of bioavailable heavy metals including mercury, arsenic, and chromium; solvents; and pesticides (including toxaphene).
- Level C personal protective equipment is required when working with blood meal.

FULL-SCALE TREATMENT EXAMPLES (REFS. 1, 2, 5 AND 6):

Anaerobic bioremediation using blood meal and phosphate amendments has been implemented at a full scale at 22 Dip Vat sites in the Navajo Nation. Other sites where this technology has been applied at a full scale to remediate toxaphene-contaminated soil include:

(1) The Ojo Caliente Dip Vat site
(2) The Laahty Family Dip Vat site
(3) The Henry O Dip Vat site
(4) The Acoma Reservation at Sky City
(5) The Gila River Indian Community (GRIC 1) crop duster site
(6) The GRIC 2 crop duster site

The resources used for this fact sheet contain performance data on nine applications of this technology. Performance data for each of these sites is presented in Table 1 at the end of this fact sheet. Three of these sites are discussed below in greater detail. The unit cost of implementation at these sites in USD ranged from $98 to $296 per cubic yard.

Laahty Family Dip Vat (LDV) site

The LDV site is located in The Zuni Nation, New Mexico. Soil at the site was contaminated with toxaphene at an average concentration of 29 milligrams per kilogram (mg/kg). A total of 253 cubic yards (cy) of soil was excavated and stockpiled on site. A cell with dimensions, 73 feet (ft) by 30 ft by 4 ft (deep), was constructed and lined with a plastic liner. Contaminated soil was placed in a concrete mixer and mixed with biological amendments and water. Blood meal and monobasic phosphate were added, each at a dosage rate of 10 grams per kilogram (g/kg) of contaminated soil. Dibasic phosphate salts were also added at a dosage rate of 3.3 g/kg soil. The nutrient-amended soil slurry was then placed in the lined cell. Water was added to provide one foot of cover above the solids in the cell. The cell was then covered with a plastic sheet and incubated. Samples were collected periodically to monitor progress. The toxaphene concentration decreased in the anaerobic cell from an initial concentration of 29 mg/kg to 4 mg/kg in 31 days. This corresponded to an overall reduction of 86 percent. The post-treatment concentrations were below the 17 mg/kg action level established for the site. In 2004, the total cost of treatment in USD was $75,000. Consequently, the unit cost of treatment at this site was $296 per cubic yard.

Henry O Dip Vat (HDV) Site

The HDV site is located in The Zuni Nation, New Mexico. Approximately 660 cy of soil at this site was contaminated with toxaphene at an average concentration of 23 mg/kg. Two cells were constructed for soil treatment:

- The north cell (Cell 1) was 75 ft by 35 ft by 5 ft (deep).
- The south cell (Cell 2) was 65 ft by 30 ft by 5 ft (deep).

Both cells were lined with plastic liners. Blood meal and sodium phosphate were added to contaminated soil and placed in a mixing pit using a backhoe. The dosage rate of blood meal was 5 g/kg of contaminated soil, while that of monobasic phosphate was 10 g/kg of contaminated soil. Dibasic phosphate salts were also added at a dosage rate of 3.3 g/kg. Water was added to the soil in the mixing pit, and the resulting soil slurry was extensively mixed. Once mixed, the soil slurry was transferred to anaerobic cells 1 and 2. Water was added to provide one foot of additional cover above the solids in each cell. Each cell was then covered with a plastic sheet and incubated for 61 to 76 days. Samples were collected on day 1 and day 61 from Cell 1 and on day 1 and 76 from Cell 2. Analysis of the samples indicated that the average toxaphene concentration was reduced from 23 mg/kg to 8 mg/kg. This corresponds to a percent removal of approximately 67 percent removal in 68 days. The post-treatment concentrations were below the 17 mg/kg action level established for the site. In 2004, the total cost of treatment in USD was $65,000. Consequently, the unit cost of treatment at this site was $98 per cubic yard.

Gila River Indian Community Site

The Gila River Indian Community (GRIC) site is located in Chandler, Arizona. Approximately 3,500 cy of toxaphene-contaminated soil required treatment at this site. Four lined cells were constructed with dimensions of 178 ft by 43 ft by 7 ft (deep). This dosage rate was lower than for other sites to reduce costs. The dosage rate of blood meal, sodium phosphate, and dibasic phosphates was 5 g/kg of contaminated soil. Blood meal and phosphates were first mixed in a pit, and then blended with contaminated soil using a pug mill (100-300 cy/hr throughput). The mixture was then transferred to cells filled with water to 25 percent capacity. Additional water was then added to the cells to provide one foot of cover above the solids. Each cell was then covered with a plastic sheet. Samples were collected from the cells after initial setup and at the end of 3 months, 6 months, and 9 months. The removal of toxaphene in GRIC site soil took longer than usual due to the reduced amendment dosage rates. The average toxaphene concentration at the end of 180 days ranged between 4 mg/kg and 5 mg/kg demonstrating 83 to 88 percent toxaphene removal. The samples collected at day 272 showed residual levels of 2 to 4 mg/kg corresponding to a percent removal between 87 and 98 percent. The post-treatment concentrations were below the 17 mg/kg action level established for the site. In 2004, the total cost of treatment in USD was $793,000. Consequently, the unit cost of treatment at this site was $226 per cubic yard.

| Table 1 Performance Data for Anaerobic Bioremediation of Toxaphene Using Blood Meal at Selected Sites | | | | | |
|---|---|---|---|---|---|
| **Site Name** | **Untreated Concentration (mg/kg)** | **Treated Concentration (mg/kg)** | **Period (Days)** | **Percent Reduction** | **Volume Treated (cy)** |
| Navajo Vats Chapter | | | | | |
| Nazlini | 291 | 71 | 108 | 76 | NA |
| Whippoorwill | 40 | 17 | 110 | 58 | NA |
| Blue Canyon Road | 100 | 17 | 106 | 83 | NA |
| Jeddito Island | 22 | 3 | 76 | 77 | NA |
| Poverty Tank | 33 | 8 | 345 | 76 | NA |
| Ojo Caliente | 14 | 4 | 14 | 71 | 200 |
| Laahty Family Dip Vat | 29 | 4 | 31 | 86 | 253 |
| Henry O Dip Vat | 23 | 8 | 68 | 67 | 660 |
| Gila River Indian Community | | | | | |
| Gila River Indian Community (Cell 1) | 59 | 4 | 272 | 94 | |
| Gila River Indian Community (Cell 2) | 31 | 4 | 272 | 87 | |
| Gila River Indian Community (Cell 3) | 29 | 2 | 272 | 94 | |
| Gila River Indian Community (Cell 4) | 211 | 3 | 272 | 98 | 3,500 |

Note:
mg/kg:       Milligrams per kilogram
NA:           Not available
Source:      Refs. 1, 2 and 6

| **U.S. EPA** CONTACT: | **LAAHTY FAMILY AND HENRY O DIP VAT SITES:** | **Gila River Indian Community** CONTACT: |
|---|---|---|
| U.S. EPA Environmental Response Team<br>Harry L. Allen III, Ph.D.<br>Phone: (732) 321-6747<br>Email: allen.harry@epa.gov | Bureau of Indian Affairs<br>Southwest Region<br>Zuni Nation<br>Phone: (505) 563-3106 | GRIC Department of Environmental Quality Hazardous Waste Program Manager<br>Dan Marsin<br>Email: hazmat@gilnet.net<br>Phone: (520) 562-2234 |

**PATENT NOTICE:**
This technology has not been patented.

**REFERENCES:**

1.  Allen L., Harry and others. 2002. Anaerobic bioremediation of toxaphene-contaminated soil – a practical solution. 17th WCCS, Symposium No. 42, Paper No. 1509, Thailand. August 14 – 21.

2.  Allen L., Harry, US EPA Environmental Response Team. 2005. Email to Younus Burhan, Tetra Tech EM Inc., Regarding Comments from Harry L. Allen on Draft (January 5, 2005) Blood Meal Fact Sheet. January 25.

3.  Allen L., Harry, US EPA Environmental Response Team. 2005. Memo to Ellen Rubin, US EPA Office of Superfund Remediation and Technology Innovation. Response to Questions on Toxaphene Fact Sheet. February 24.

4.  U.S. Environmental Protection Agency (US EPA). Office of Superfund Remediation and Technology Innovation. 2004. Cost and Performance Summary Report. The Legacy of the Navajo Vats Superfund Site, Arizona and New Mexico. October.

5.  US EPA. 2000. Fact Sheet - Gila River Indian Community Toxaphene Site. October.

6.  Rubin, Ellen, US EPA Environmental Response Team. 2005. Email to Younus Burhan, Tetra Tech EM Inc., Regarding Comments from Dr. T. Ferrell Miller on Draft (January 5, 2005) Blood Meal Fact Sheet. February 7.

# APPENDIX C

# Fact Sheet on
# Bioremediation Using DARAMEND® for Treatment of
# POPs in Soils and Sediments

**POPs - WASTES APPLICABILITY (REFS. 1, 6, AND 10):**
DARAMEND® is a bioremediation technology that has been used to treat soils and sediments containing low concentrations of pesticides such as toxaphene and DDT as well as other contaminants.

| | |
|---|---|
| **POPs Treated:** | Toxaphene and DDT |
| **Other Contaminants Treated:** | DDD, DDE, RDX, HMX, DNT, and TNT |

**TECHNOLOGY DESCRIPTION (REFS. 4, 5 AND 10):**
**OVERVIEW**
DARAMEND® is an amendment-enhanced bioremediation technology for the treatment of POPs that involves the creation of sequential anoxic and oxic conditions. The treatment process involves the following:

1. Addition of solid phase DARAMEND® organic soil amendment of specific particle size distribution and nutrient profile, zero valent iron, and water to produce anoxic conditions.
2. Periodic tilling of the soil to promote oxic conditions.
3. Repetition of the anoxic-oxic cycle until the desired cleanup goals are achieved.

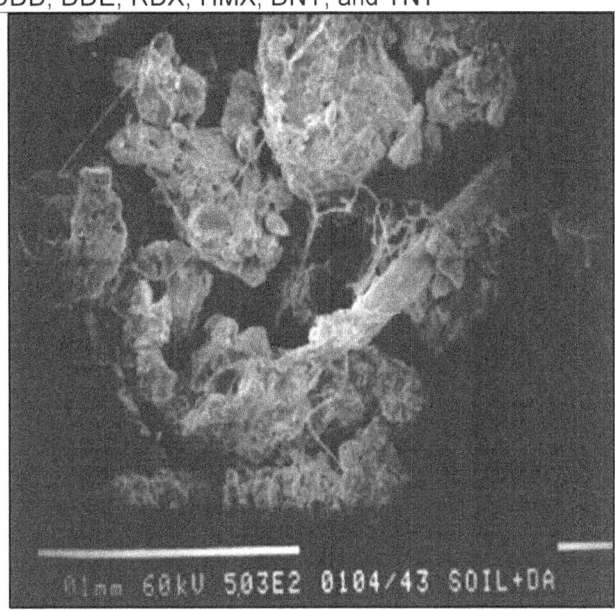

DARAMEND® particle colonization as viewed through an electron-microscope
Source: Adventus Americas, Inc.

The addition of DARAMEND® organic amendment, zero valent iron, and water stimulates the biological depletion of oxygen generating strong reducing (anoxic) conditions within the soil matrix. The diffusion of replacement oxygen into the soil matrix is prevented by near saturation of the soil pores with water. The depletion of oxygen creates a very low redox potential, which promotes dechlorination of organochlorine compounds. A cover may be used to control the moisture content, increase the temperature of the soil matrix and eliminate run-on/run off. The soil matrix consisting of contaminated soil and the amendments is left undisturbed for the duration of the anoxic phase of treatment cycle (typically 1- 2 weeks).

In the oxic phase of each cycle, periodic tilling of the soil increases diffusion of oxygen to microsites and distribution of irrigation water in the soil. The dechlorination products formed during the anoxic degradation process are subsequently removed trough aerobic (oxic) biodegradation processes, initiated by the passive air drying and tilling of the soil to promote aerobic conditions.

Addition of DARAMEND® and the anoxic-oxic cycle continues until the desired cleanup goals are achieved. The frequency of irrigation is determined by weekly monitoring of soil moisture conditions. Soil moisture is maintained within a specific range below its water

Application of DARAMEND® at the T.H. Agricultural and Nutrition Superfund Site (Source: Adventus Americas, Inc.).

holding capacity. Maintenance of soil moisture content within a specified range facilitates rapid growth of an active microbial population and prevents the generation of leachate. The amount of DARAMEND® added in the second and subsequent treatment cycles is generally less than the amount added during the first cycle.

DARAMEND® technology can be implemented using land farming practices either ex situ or in situ. In both cases, the treatment layer is 2 feet (ft) deep, the typical depth reached by tilling equipment. However, the technology can be implementation in 2-ft sequential lifts. In the ex situ process, the contaminated soil is excavated and sometimes mechanically screened in order to remove debris that may interfere with the distribution of the organic amendment. The screened soil is transported to the treatment unit, which is typically an earthen or concrete cell lined with a high-density polyethylene liner. In situ, the soil may be screened to a depth of 2-ft using equipment such as subsurface combs and agricultural rock pickers.

**STATUS AND AVAILABILITY (REF. 1):**
DARAMEND® is a proprietary technology and is available only through one vendor - Adventus Remediation Technologies (ART), Mississauga, Ontario, Canada. In the U.S., the technology is provided by ART's sister company, Adventus Americas Inc., Bloomingdale, IL. The technology has been used for the treatment of POPs (toxaphene and DDT) since 2001. Table 1 lists performance data for DARAMEND® technology application at selected sites. Through 2005, DARAMEND® has been implemented at two POPs contaminated sites.

| | | | | | | Table 1: Performance Data of DARAMEND at Selected Sites | | | |
|---|---|---|---|---|---|---|---|---|---|
| | | | | | | | Performance | | |
| Site Name | Scale | Quantity Treated (tons) | No. of treatment cycles | Duration of each cycle | Cost per ton* | Contaminant | Untreated Concentration (mg/kg) | Treated Concentration (mg/kg) |
| **POPs Contaminated Sites** | | | | | | | | | |
| T.H. Agricultural & Nutrition (THAN) Superfund Site, Montgomery, Alabama | Full | 4,500 | 15 | 10 days | $55 | Toxaphene DDT DDE DDD | See Table 2 for performance data | |
| W.R. Grace, Charleston, South Carolina | Pilot | 250 | 8 | 1 month | $95 | Toxaphene | 239 | 5.1 |
| | | | | | | DDT | 89.7 | 16.5 |
| **Non-POPs Contaminated Sites** | | | | | | | | | |
| Naval Weapons Station, Yorktown, Virginia | Full | 4,800 | 12 | 7-10 days | $90 | TNT | 15,359 | 14 |
| | | | | | | RDX | 1,090 | 1.6 |
| | | | | | | DNT | 1,002 | 13 |
| Iowa Army Ammunition Plant, Burlington, Iowa | Full | 8,000 | 5 | 7-10 days | $150 | RDX | 1,530, | 16.2 |
| | | | | | | HMX | 1,112, | 84.5 |
| | | | | | | TNT | 95.8 | 8 |
| Confidential Site, Northwest U.S.A. (applied in multiple 2-ft lifts) | Full | 6,000 | Aerobic treatment | NA | $37 | PCP | 359 | 8 |
| | | | | | | PCP | 760 | 31 |

*Source: Ref. 1*
*\* Treatment costs are as reported by vendor. The vendor did not specify what was included in this cost.*

**DESIGN (REF. 5):**
The major design factor for the implementation of this technology is the amount and type of soil amendments required for bioremediation. This is dependent on site conditions and the physical (textural variation, percent organic matter, and moisture content) and chemical (soil pH, macro and micronutrients, metals, concentration and nature of contaminants of concern) properties of the target soil. The duration of the treatment cycle is based on soil chemistry, concentration of contaminants of concern and soil temperature. The number of treatment cycles is based on the required cleanup levels of the contaminant.

**THROUGHPUT (REF. 4):**
For ex situ treatment, the amount of POPs contaminated soil/sediment that can be treated is dependent on the available surface area to spread contaminated soil. The technology can also be applied ex situ in windrows. For in situ application, the tillage equipment limits the depth (2-ft) to which the soil can be remediated. However, the technology can be used in situ at depth greater than 2-ft using alternative soil mixing equipment or injection techniques.

**WASTES/RESIDUALS (REF. 4):**
The primary wastes generated are debris, stone, and construction material that are removed in the pretreatment process. No leachate is generated if a treatment area cover is used. If no cover is used, precipitation in the treatment area may generate leachate or storm water run-off.

Sampling and monitoring activities of the treatment pile will generate personal protective equipment (PPE) and contaminated water from decontamination activities.

**MAINTENANCE:**
Implementation of the DARAMEND® technology to treat POPs requires limited maintenance such as the upkeep of tilling, soil moisture control, and other industrial equipment. Because the specific amendments and application rate of DARAMEND® are site and soil-specific, the ongoing maintenance will vary by site and type of soil treated.

**LIMITATIONS (REFS. 4 AND 9):**
DARAMEND® technology may become technically or economically infeasible when treating soils with excessively high contaminant concentration. The technology has not been used for the treatment of other POPs such as PCBs, dioxins, or furans. ART, the developer of the technology, indicated that it has been only marginally successful in bench scale treatment of PCB-contaminated soil. Bench scale or pilot scale studies are typically conducted before field application of this technology; the type and amount of soil amendments required are then based on the results of these studies.

In situ application of this technology using tilling equipment is limited to a depth of 2-ft. However, the technology can be used in situ at depths greater than 2-ft using alternative soil mixing equipment or injection techniques. This technology requires that the treatment area be free of surface and subsurface obstructions that would interfere with the soil tilling. Ex situ application of this technology requires a large surface area to treat large quantities of the contaminated soil. Implementation of this technology in 2-ft sequential lifts would increase the total time required to treat the contaminated soil. The technology can also be applied ex situ in windrows.

Application of this technology requires a source of water (either city, surface, or subsurface).

This technology cannot be applied to sites that are prone to seasonal flooding or have a water table that fluctuates to within 3-ft of the site surface. These conditions make it difficult to maintain the appropriate range of soil moisture required for effective bioremediation, and may redistribute contamination across the site.

Volatile organic compound emissions may increase during soil tilling. Other factors that could interfere with the process would be large amounts of debris in the soil, which would interfere with the incorporation of organic amendments and reduce the effectiveness of tilling. Presence of other toxic compounds (heavy metals) may be detrimental to soil microbes. Soils with high humic content may slow down the cleanup through increased organic adsorption and oxygen demand.

**FULL-SCALE TREATMENT EXAMPLES (REF. 3):**
Bioremediation of pesticides-impacted soil/sediment, T.H. Agriculture and Nutrition (THAN) Superfund Site, Montgomery, Alabama.

The THAN site is located on the west side of Montgomery, Alabama, about 2 miles south of the Alabama River. The site is approximately 16 acres in area. Previous site operations involved the formulation, packing and distribution of pesticides, herbicides, and other industrial/waste treatment chemicals. The site was listed on the National Priorities List (NPL) on August 30, 1990. In 1991, US EPA entered into a consent agreement with Elf Atochem North America Inc., the Potentially Responsible Party (PRP) for the site, to conduct a remedial investigation/feasibility study for the site. The final Record of Decision (ROD) for the site was signed on September 28, 1998, and bioremediation was selected as the remedy for treating the contaminated soils and sediments. DARAMEND® was selected as the bioremediation technology.

The contaminated soil and excavated sediments (approximately 4,500 tons) were treated using anaerobic/aerobic bioremediation cycle using DARAMEND®. Implementation of the technology involved the following steps:

1. DARAMEND® amendment and powdered iron application and incorporation
2. Determination of water holding capacity (first cycle only)
3. Determination of treatment matrix moisture content
4. Irrigation
5. Measurement of soil redox potential
6. Soil allowed to stand undisturbed for anoxic phase (approximately 7 days)
7. Soil tilled daily to generate oxic condition (approximately 4 days)
8. Steps 1, and 3 to 7 were repeated for each subsequent cycle. Fifteen treatment cycles were implemented in some treatment areas on site.

Two agricultural tractors (Model: Massey-Ferguson 394 H) mounted with deep rotary tillers were used for amendment application and tilling the treatment area. The target soil moisture content at the beginning of each cycle was approximately 33% (dry wt. basis) or 90% of the soil's water holding capacity. The optimal pH range (6.6 to 8.5) of the treatment area was maintained by adding hydrated lime at a rate of 1,000 mg/kg during the oxic phase of the third, sixth, and twelfth cycle. Following the application of each treatment cycle, samples were collected from the treatment area. The treatment area was divided into 12 sampling zones and one composite sample (composite of four grab samples) was collected from each zone. The samples were collected from the full 2-ft soil profile of treatment area. Fifteen treatment cycles were applied to some areas of the site. Table 2 lists the initial and final concentration of the samples collected from these 12 zones.

Based on the final sampling event DARAMEND® reduced the concentration of all the contaminants of concern to less than the specified performance standards. The average treatment cost in USD at the THAN site was $55 per ton. The vendor did not specify what was included in this cost.

| Sampling Zone | Toxaphene (29 mg/kg) [1] | | DDT (94 mg/kg) [1] | | DDD (94 mg/kg) [1] | | DDE (133 mg/kg) [1] | |
|---|---|---|---|---|---|---|---|---|
| | Initial [2] Conc. (mg/kg) | Final [3] Conc. (mg/kg) | Initial [2] Conc. (mg/kg) | Final [3] Conc. (mg/kg) | Initial [2] Conc. (mg/kg) | Final [3] Conc. (mg/kg) | Initial [2] Conc. (mg/kg) | Final [3] Conc. (mg/kg) |
| 1 | 77 | < 20 | 126 | 10.2 | 52 | 26.4 | 33 | 6 |
| 2 | 260 | < 21 | 227 | 15 | 133 | 73 | 35.3 | 8.4 |
| 3 | 340 | < 21 | 33.2 | 4.5 | 500 | 89 | 49 | 7.8 |
| 4 | 45 | < 21 | 55.1 | 14.7 | 34 | 37 | 15.8 | 7.2 |
| 5 | 230 | < 21 | 216 | 16.1 | 93 | 53 | 22.4 | 6.8 |
| 6 | 90 | < 21 | 13.3 | 2.2 | 130 | 59 | 17 | 5.7 |
| 7 | 100 | < 20 | 151 | 15.3 | 85 | 38 | 25.2 | 6.3 |
| 8 | 13 | < 20 | 9.1 | 5.2 | 44 | 24.3 | 6.9 | 2.8 |
| 9 | 330 | < 21 | 45 | 5.7 | 312 | 85 | 28.2 | 7.2 |
| 10 | 48 | < 20 | 44.4 | 5.7 | 146 | 25.5 | 20.1 | 4.2 |
| 11 | 20 | < 20 | 12.6 | 2.9 | 46 | 25.1 | 6.9 | 3.0 |
| 12 | 720 | < 21 | 78 | 6.3 | 590 | 87 | 59.6 | 8.6 |

**Table 2: DARAMEND® performance at the THAN Site**

Notes:
1. Performance Standard as specified in the Record of Decision, Summary of Remedial Alternatives Selection, THAN Site.
2. Initial concentration reported from samples collected by responsible party.
3. Final concentration reported from splits samples collected by US EPA.

**U.S. EPA RPM FOR THAN SITE:**
Brian Farrier
EPA Region 4
Telephone: 404-562-8952
Fax:          404-562-8955
Email:       farrier.brian@epa.gov

**VENDOR CONTACT DETAILS:**
David Raymond
Adventus Remediation Technologies, Inc.
1345 Fewster Drive
Mississauga, Ontario L4W 2A5
Telephone:  905-273-5374, Extension 224
Mobile:        416-818-0328
Fax:            905-273-4367
Email: info@AdventusGroup.com
Website: http://www.adventusgroup.com/

**PATENT NOTICE:**
DARAMEND® is a patented technology with U.S. Patent No. 5,618,427.

REFERENCES:

1. Adventus Remediation Technologies, Inc. DARAMEND project summaries. Online Address: http://www.adventusremediation.com.

2. Adventus Remediation Technologies, Inc. March 2002. Draft Final Report, Ex-Situ DARAMEND Bioremediation of Soil Containing Organic Explosive Compounds, Iowa Army Ammunition Plant, Middletown, Iowa.

3. Adventus Remediation Technologies, Inc. November 2003. Final Report, Bioremediation of Soil and Sediment Containing Chlorinated Organic Pesticides, THAN Superfund Site, Montgomery, Alabama.

4. US EPA. 1996. Site Technology Capsule, GRACE Bioremediation Technologies DARAMEND® Bioremediation technology. Superfund Innovative Technology Evaluation. EPA/540/R-95/536.

5. US EPA. 1997. Site Technology Capsule, GRACE Bioremediation Technologies DARAMEND® Bioremediation technology. Superfund Innovative Technology Evaluation. EPA/540/R-95/536a.

6. US EPA. 2002. Technology News and Trends, Full-Scale Bioremediation of Organic Explosive contaminated soil. EPA 542-N-02-003. July.

7. US EPA. 2004. TH Agricultural & Nutrition Company Site Information and Source Data. Online Address: http://www.epareachit.org.

8. US EPA. 2004. TH Agricultural & Nutrition Company Site, RODS Abstract information, Superfund Information Systems. Online Address: http://www.epa.gov/superfund.

9. Farrier, Brian, US EPA Region 4. 2004. Telephone Conversation with Younus Burhan, Tetra Tech EM Inc. August 31 and October 19.

10. Phillips, T., Bell, G., Raymond, D., Shaw, K., and Seech, A. 2001. "DARAMEND® technology for in situ bioremediation of soil containing organochlorine pesticides."

11. Raymond, David, Adventus Remediation Technologies, Inc. 2004. Telephone Conversation with Younus Burhan, Tetra Tech EM Inc. August 25.

# APPENDIX D

## Fact Sheet on
## In Situ Thermal Desorption for Treatment of POPs in
## Soils and Sediments

**POPs-WASTES APPLICABILITY (REFS. 4 AND 16):**

ISTD is a thermally enhanced in Situ treatment technology that uses conductive heating elements to directly transfer heat to environmental media. ISTD can heat soil or sediment in situ to average temperatures of 1,000 degrees Fahrenheit (°F), and as a result has been used to treat compounds with relatively high boiling points. Some of these include semivolatile organic contaminants (SVOCs) such as polychlorinated biphenyls (PCBs), polycyclic aromatic hydrocarbons (PAHs), pesticides, and herbicides. Pilot- and full-scale applications have been performed where ISTD has been used to remove PCBs, and where dioxins and furans were trace contaminants. TerraTherm is the sole vendor for ISTD. According to TerraTherm, laboratory-scale work and extrapolation techniques have suggested the potential applicability of ISTD to POPs other than PCBs, dioxins, and furans (including aldrin, dieldrin, endrin, chlordane, heptachlor, DDT, mirex, hexachlorobenzene, and toxaphene); however, these contaminants have not yet been treated using ISTD on a full- or pilot-scale basis. ISTD has been used to treat contaminants in most hydrogeologic settings, including beneath structures.

| | |
|---|---|
| **POPs Treated:** | PCBs, dioxins, and furans, aldrin, chlordane, dieldrin, and endrin |
| **Other Contaminants Treated:** | Hexachlorocyclopentadiene, isodrin, VOCs, SVOCs, oils, creosotes, coal tar PAHs, gasoline and diesel range organics, and MTBE |

**TECHNOLOGY DESCRIPTION (REFS. 2, 4, 13 AND 16):**

**OVERVIEW**

ISTD involves simultaneous application of heat and vacuum to subsurface soils. There are three basic elements in an ISTD process: (1) application of heat to contaminated media; (2) collection of desorbed contaminants through vapor extraction; and (3) treatment of collected vapors. Figure 1 presents a typical ISTD system.

ISTD has been used at full scale to treat PCBs, PAHs, dioxins, and chlorinated volatile organic compounds (CVOC). At the temperatures achieved by the ISTD process, volatiles metals such as mercury may also be recovered.

Figure 1
Typical ISTD System

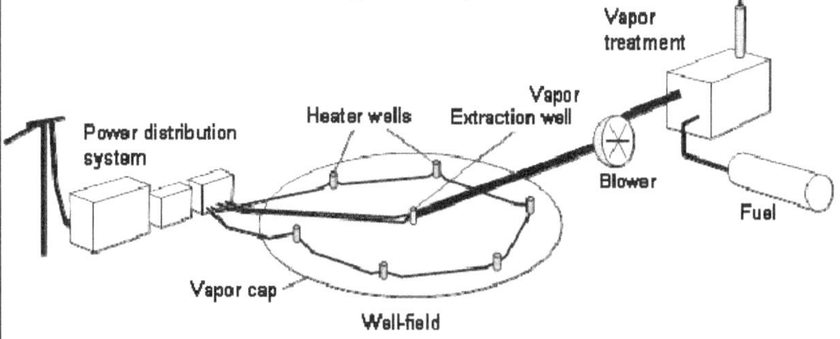

Source: TerraTherm™ Inc.

**In Situ Heating**

ISTD uses surface or buried electrically powered heaters to heat contaminated media. The most common setup uses a vertical array of heaters placed inside wells drilled into the remediation zone. A less common setup uses the same type of heaters installed horizontally on the surface of the contaminated zone. This method of heating (often called blanket heating) is typically used when contamination is shallow (usually 1 to 3 feet below ground surface (bgs)). Figure 2 illustrates the two different methods of heating.

ISTD heaters can attain temperatures as high as 1,600°F, and can produce average media temperatures exceeding 1,000°F. Heat originates from a heating element and is transferred to the subsurface largely via thermal conduction and radiant heat transport, which dominates near the heat sources. There is also a contribution through convective heat transfer that occurs during the formation of steam from pore water present in the soil or sediment.

The thermal conductivity values of a wide range of soil types (e.g., clay, silt, sand, gravel) vary only by a factor of approximately four. Therefore, the rate of heat transfer from the linear heaters to the surrounding media is radially uniform. When heating commences, the temperature profile in the remediation zone is characterized by large gradients, and temperatures decrease sharply with distance from the source. Over time, superposition of heat from adjacent heaters tends to even out these differences.

**Figure 2**
**Blanket and Thermal Well Heating**

Blanket
Heating

Thermal Well
Heating

Source: TerraTherm™ Inc.

## Vapor Extraction

As the matrix is heated, adsorbed and liquid-phase contaminants begin to vaporize. A significant portion of organic contaminants either oxidize (if sufficient air is present) or pyrolize once high soil temperatures are achieved. Desorbed contaminants are recovered through a network of vapor-extraction wells.

Vapor extraction wells are also heated to prevent condensation of contaminants inside the well. A vacuum is applied to these wells to induce air flow through the contaminated media creating a zone of capture. Contaminant vapors captured by the extraction wells are conveyed to an offgas treatment system for treatment prior to discharge to the atmosphere.

## Offgas Treatment (Ref. 2)

TerraTherm offers two different methods of vapor treatment. One treats extracted vapor without phase separation (Figure 1), and the other cools heated vapor, separates the resulting phases, and manages each phase separately.

The vapor treatment option depicted by Figure 1 uses a thermal oxidizer to break down organic vapors to primarily carbon dioxide and water. Stack sampling has demonstrated that toxic pollutants in offgas, including dioxins, are substantially below regulatory standards. When influent vapors contain chlorinated compounds, hydrogen chloride (HCl) gas is produced. In such cases, the exhaust from the thermal oxidizer is passed through an acid gas scrubber to capture HCl gas.

The other vapor treatment option uses a heat exchanger to cool extracted vapors. The resulting liquid phase is then separated into aqueous and nonaqueous phases. The nonaqueous phase liquid (NAPL) is usually disposed of at a licensed treatment storage and disposal facility. The aqueous phase is passed through liquid-phase activated carbon adsorption units and then released into the environment. Cooled, uncondensed vapor is passed through vapor-phase activated carbon adsorption units and then vented to atmosphere.

Although setup varies from site to site, several components of the remediation system including heaters, blowers, and offgas treatment equipment are either standard or adaptable to new situations, with equipment reused from site to site. Downhole wells may not be salvageable and may be plugged and abandoned in place.

**STATUS AND AVAILABILITY (REFS. 4 AND 5):**
ISTD is a patented technology originally developed by Shell Oil. While U.S. Patent rights were donated to the University of Texas (UT), patent rights outside the U.S. were retained by Shell. TerraTherm holds the exclusive license to this technology from both UT and Shell, and is currently the only vendor. ISTD has been commercial for several years. Its ability to remove PCBs from contaminated soil was first demonstrated more than 6 years ago. As shown on Table 1, ISTD has been used at six POP-contaminated sites. Implementation at four of these sites was full scale, and the other two were pilot scale.

**Table 1**
**Performance of ISTD at POPs Contaminated Sites (Refs. 2, 4 and 7)**

| Site Name | Year | Scale | Contaminant | Concentration | | | |
| --- | --- | --- | --- | --- | --- | --- | --- |
| | | | | Initial | Final | Goal | Units |
| Former South Glens Falls Dragstrip, Moreau, New York | 1996 | Full | PCB 1248/1254 | 5,000 (Max) | < 0.8 | 2 | mg/kg |
| Tanapag Village, Saipan, NMI | 1997 - 1998 | Full | PCB 1254/1260 | 10,000 (Max) | < 1 | 10 | mg/kg |
| Centerville Beach, Ferndale, CA | 1998 - 1999 | Full | PCB 1254 | 860 (Max) | < 0.17 | 1 | mg/kg |
| | | | Dioxins and Furans | 3.2 (Max) | 0.006 [1] | 1 | ug/kg TCDD |
| Missouri Electric Works, Cape Girardeau, MO | 1997 | Pilot | PCB 1260 | 20,000 (Max) | < 0.033 | 2 | mg/kg |
| Former Mare Island Naval Shipyard, Vallejo, CA | 1997 | Pilot | PCB 1254/1260 | 2,200 (Max) | < 0.033 | 1 | mg/kg |
| Former Wood Treatment Area, Alhambra, CA | 2002 - 2005 | Full | Dioxins and Furans | 18 (Mean) | 0.01 | 1 | ug/kg |

Note:

| | |
| --- | --- |
| Avg | Average concentration |
| Max | Maximum concentration |
| mg/kg | Milligrams per kilogram (or parts per million) |
| NMI | Northern Mariana Islands |
| ND | Below detection limit |
| TCDD | Tetrachlorodibenzodioxin equivalents |
| ug/kg | Micrograms per kilogram (or parts per billion) |

[1]     Final concentration presented as average of residual concentrations in treatment area.

**DESIGN (REF. 12):**
Key design factors for ISTD include the number and depth of heater wells and vacuum wells, as well as the requirements for electrical power and treatment of off gasses. These factors are affected by the type of contaminants present, concentration of the contaminants, extent of contamination, soil type, hydraulic conductivity, permeability, thermal properties, location of the water table, availability of site facilities such as electrical power supply, and regulatory issues.

THROUGHPUT (REF. 5):
ISTD has been used to treat volumes as low as a few hundred cubic yards to greater than 20,000 cubic yards in 6 to 9 months. Factors affecting cleanup durations can include type of contaminants, cleanup/remedial goals, and site geology.

WASTES/RESIDUALS (REFS. 3 AND 5):
Wastes produced by ISTD are likely to result from the treatment of extracted vapors, and vary according to the type of treatment they are subjected to. Offgas treatment options that employ phase separation techniques could produce process wastes such as NAPL, spent liquid- and vapor-phase activated carbon, and inorganic salts as waste products. For example, the treatment of chlorinated vapors in a thermal oxidizer results in the production of HCL gas. A wet or dry acid gas scrubber used to neutralize HCl gas will produce inorganic salts as a waste product.

NAPL is typically transported off site for disposal at a licensed facility. Spent activated carbon may either be disposed of, or regenerated at a licensed facility. Inorganic salts produced from neutralization processes are typically considered nonhazardous and are consequently disposed of as nonhazardous waste.

MAINTENANCE (REF. 4):
Maintenance associated with ISTD includes the occasional replacement of heater elements. ISTD operation is typically characterized by less than 5% downtime. Other maintenance needs include treatment media replacement and thermal oxidizer refueling.

LIMITATIONS (REF. 4):
The following are some of the limitations of this technology:

- ISTD cannot address contaminants that do not volatilize with in the temperature range of approximately 15-1000°C.
- As long as liquid water remains within the remediation zone, the temperature that can be attained is limited to the boiling point of water (212°F). Once the water is boiled off, higher temperatures can be attained. A continuing source of water influx into the treatment zone will undermine the ability of this technology to produce temperatures necessary for the removal of POPs. For this reason, formation dewatering and implementation of water control measures are needed prior to the implementation of ISTD in high-permeability, water-saturated zones.
- Though not always the case, cost can be a limiting factor. Unit costs for treatment are influenced by several factors including scale of the project, depth of the treatment zone, depth to water table, air emission controls, cost of labor and cost of power. However, in general, unit costs in USD range from $200 to $600 per cubic yard corresponding to treatment volumes ranging from less than 5,000 to approximately 15,000 cubic yards for POP-type contaminants. Larger volumes may have lower unit costs. Treatment costs for VOC contaminants are lower.

FULL-SCALE TREATMENT EXAMPLES:

Centerville Beach (Refs. 6, 8, 10 and 14)

The Centerville Beach Naval Facility is a 30-acre site in Ferndale, California that was used for oceanographic research and undersea surveillance. The site was decommissioned in 1993. Operations at the site lead to contamination of a particular area with PCBs. The PCB of concern was Aroclor 1254 which was present in concentrations ranging from 0.15 to 860 milligrams per kilogram (mg/kg). Dioxins and furans were also present at a maximum concentration of 3.2 micrograms per kilogram (µg/kg) as 2,3,7,8-tetrachlorodibenzodioxin (TCDD) equivalents. The contaminated medium was primarily silty clay. Groundwater was encountered below the contaminated zone at depths exceeding 60 feet bgs.

From September 1998 through February 1999, approximately 1,000 cubic yards of PCB-contaminated soil was treated using ISTD. Heater and vapor extraction wells were installed in a zone measuring 40 feet long, 30 feet wide, and 15 feet deep. The wells were installed 6 feet apart. Two sealed vacuum blowers were used in parallel for vapor extraction. Offgas was treated using a flameless thermal oxidizer (with greater than 99.99% demonstrated treatment efficiency), and two granular activated carbon units configured in series. The total cost of the implementation in USD was approximately $650,000.

The treatment goal was 1 mg/kg for PCBs and 1 µg/kg TCDD equivalent for dioxins and furans. Remediation took place between September 1998 and February 1999. Treatment goals were met in the bulk of the treatment area; however, one portion (178 cubic yards) still contained elevated concentrations of PCBs. This was found to be caused by a previously undiscovered bank of PCB-containing electrical conduits emanating from outside the treatment zone and passed into the treatment area. Excavation and disposal was subsequently used to remove this area of contaminated soil and the associated conduits.

Alhambra (Refs. 3, 9, 17 and 18)

Southern California Edison's (SCE) Alhambra Combined Facility occupies approximately 33 acres and is currently used for storage, maintenance, and employee training. SCE carried out wood treatment operations in SCE's 2-acre former wood treatment area between 1921 and 1957. The total volume of contaminated soil was estimated to be 16,200 cubic yards of soil. The contaminated zone included a variety of buried features including treatment tanks, the structural remains of the former boiler house and tank farm, and various buried utilities. The contaminants of concern were PAHs, pentachlorophenol (PCP), and dioxins. Total PAHs were present in site soils at a maximum concentration of 35,000 mg/kg and an average concentration of 2,306 mg/kg. PCP was present at a maximum concentration of 58 mg/kg and an average concentration of less than 1 mg/kg. Dioxins were present at a maximum concentration of 0.194 mg/kg and an average concentration of 0.018 mg/kg (expressed as 2,3,7,8-tetrachloro-dibenzodioxin [TCDD] Toxic Equivalency Quotient [TEQ]). The soil in the remediation zone was composed of silty sands, inter-bedded with sands, silts, and clays. The average thermal treatment depth was approximately 20 feet bgs and extended to 100 feet bgs in some areas. The depth to the water table was greater than 240 feet bgs. The treatment goals were 0.065 mg/kg (expressed as benzo(a)pyrene [B(a)P] toxic equivalents) for PAHs; 2.5 mg/kg for PCP, and 0.001 mg/kg for dioxins (expressed as TEQ).

Remedial action at the site was conducted in two phases. Each phase addressed a different area of the site. The overall ISTD system for the two phases consisted of 785 thermal wells (131 heater-vacuum and 654 heater-only wells) at a 7.0-ft spacing between thermal wells, as well as an insulated surface seal, thermal oxidizer, heat exchanger, and granular activated carbon for off-gas treatment.

The ISTD began cleanup operations for Phase I of the remediation of Area of Concern (AOC)-2 in February 2003.

Confirmation soil samples were submitted to DTSC in July 2004 which confirmed that the cleanup goals for Phase I of AOC-2 had been achieved. Phase 2 of the cleanup began in July 2004 and was scheduled for completion by October 2004. However, a previously undiscovered

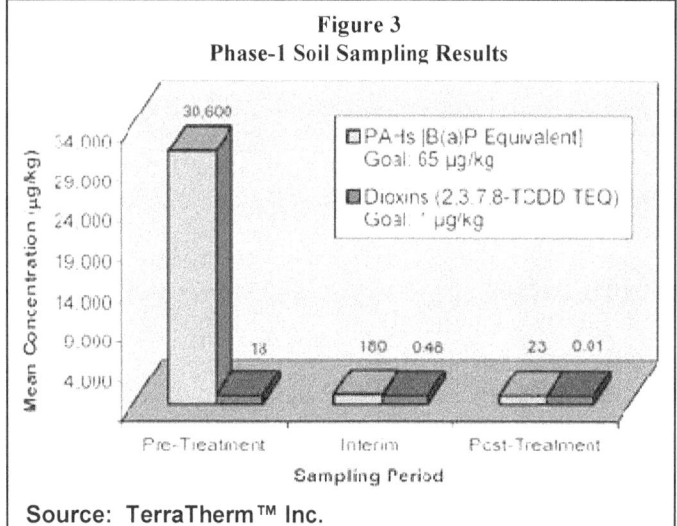

Figure 3
Phase-1 Soil Sampling Results

Source: TerraTherm™ Inc.

volume of free product made it necessary to reduce in Situ temperatures in order to control organic contaminant concentrations in the offgas treatment system influent. This resulted in an anticipated 10-month increase in the cleanup duration. Phase 2 of the cleanup was completed in August 2005. The results showed a 99% decrease in contaminant concentration from 18 ug/kg initially to .01 ug/kg. In February 2007, the Department of Toxic Substances Control (DTSC) issued a letter of closure to the

SCE Alhambra Combined Facility that states the site needed "No further action." The total cost of the implementation in USD was approximately $10 million.

Rocky Mountain Arsenal Hex Pit (Ref. 15)

The Hex Pit was a former disposal pit at the U.S. Department of Army's Rocky Mountain Arsenal. Shell Oil Company leased a portion of the Rocky Mountain Arsenal from 1952 to 1982 to manufacture pesticides. The pit was used from 1947 to 1975 to dispose of residues from distillation and other processes used in the production of hexachlorocyclopentadiene (hex), an ingredient in the manufacture of pesticides.

The main part of the Hex Pit measured approximately 94 ft by 45 ft, and varied from 8 to 10 ft deep. The pit contained a total of 2,005 cubic yards of waste-contaminated materials, of which 833 cubic yards was estimated to be waste.

The Hex Pit consisted primarily of soil and waste material originally disposed of in the pit. The impacted soil (silty sand) was stained dark brown, rust orange, or black, and at times included granules or globules of hex. Black, tar-like, relatively pure hex residue occurred in distinct solid layers of waste (approximately 1-foot thick). Hex was not detected in groundwater downgradient of the Hex Pit boundaries.

The contaminants of concern were hex, aldrin, chlordane, dieldrin, endrin, and isodrin. Only hex, chlordane, and dieldrin had treatment goals. The treatment goals were 760 mg/kg, 67 mg/kg and 335 mg/kg respectively. Laboratory tests indicated that Hex Pit wastes could be effectively treated by the ISTD process.

ISTD at the Hex Pit was designed to heat a treatment soil volume of 3,198 cubic yards, extending from 0 to 12 ft bgs and 5 ft laterally beyond the boundaries of the Hex Pit. Thermal wells on 6-foot centers were installed in a hexagonal arrangement. A total of 266 wells were installed, of which 210 were heater-only and 56 were heater-vacuum wells.

The target treatment temperature based on the boiling point of COCs was 325°C. All heater-only wells reached their operating temperatures in early March 2002. Treatment was expected to last 85 days and end in May 2002. However, 12 days after commencement, corrosion was observed in some of the well manifolds. Subsequent investigation and assessment determined that unforeseen concentration of HCL gas and production of HCL (liquid) in the vapor conveyance system, resulting from the highly concentrated wastes in the Hex Pit, had caused corrosion. Corrosion damage to the ISTD system was significant. A determination was made that replacements with necessary corrosion resisting matrices was cost prohibitive. Wastes were excavated and capped.

| STATE CONTACT (CENTERVILLE BEACH): | STATE CONTACT (ALHAMBRA): | VENDOR CONTACT: |
|---|---|---|
| California EPA Dept. of Toxic Substances Control (DTSC) Ms. Christine Parent Phone: (916) 255-3707 Email: CParent@dtsc.ca.gov | California EPA DTSC Mr. Tedd E. Yargeau Phone: (818) 551-2864 Email: tyargeau@dtsc.ca.gov | Mr. Ralph Baker TerraTherm™, Inc. Tel: (978) 343-0300 Email: rbaker@terratherm.com |

**PATENT NOTICE:**
ISTD is covered by a total of 22 U.S. patents, with 6 patents pending. TerraTherm is the exclusive licensee through the University of Texas and Shell.

**REFERENCES:**

1. Baker, Ralph and Kuhlman, Myron. 2002. 2nd International Conf. on Oxidation and Reduction Technologies for Soil and Groundwater, ORTs-2, Toronto, Ontario, Canada. A Description of the Mechanisms of In Situ Thermal Destruction (ISTD) Reactions. Nov. 17-21

2. Baker, Ralph, TerraTherm, Inc. 2004. Email to Chitranjan Christian, Tetra Tech EM Inc., Regarding Questions on ISTD. October 27, November 8, 15, 24 and 29.

3. Baker, Ralph, TerraTherm, Inc. 2004. Telephone Conversation with Chitranjan Christian, Tetra Tech EM Inc., Regarding Questions on ISTD. October 29.

4. Heron, Gorm, TerraTherm, Inc. 2004. Email to Chitranjan Christian, Tetra Tech EM Inc., Regarding Questions on ISTD. October 15.

5. Heron, Gorm, TerraTherm, Inc. 2004. Telephone Conversation with Chitranjan Christian, Tetra Tech EM Inc., Regarding Questions on ISTD. October 15.

6. Parent, Christine, California EPA, DTSC. 2004. Telephone Conversation with Chitranjan Christian, Tetra Tech EM Inc., Regarding Questions on ISTD implementation at Centerville Beach. November 2.

7. Stegemeier, G.L., and Vinegar, H.J. 2001. "Thermal Conduction Heating for In Situ Thermal Desorption of Soils," Chapter 4.6, pages 1-37. Chang H. Oh (ed.), Hazardous and Radioactive Waste Treatment Technologies Handbook, CRC Press, Boca Raton, FL.

8. TerraTherm Environmental Services. 1999. Naval Facility Centerville Beach, Technology Demonstration Report, In Situ Thermal Desorption. November.

9. TerraTherm Inc. Case Study – Alhambra. Online Address: http://www.terratherm.com/CaseStudies/WS%20Final%20Alhambra%20Sheet.pdf.

10. TerraTherm Inc. Case Study – Centerville Beach Naval Facility. Online Address: http://www.terratherm.com/CaseStudies/WS%20Centrvll-Tesi.pdf.

11. TerraTherm Inc. Case Study – Former Mare Island Naval Shipyard. Online Address: http://www.terratherm.com/CaseStudies/WS%20BADCAT.pdf.

12. TerraTherm Inc. Feasibility Screening. Online Address: http://www.terratherm.com/default.htm.

13. TerraTherm Inc. ISTD Process Description. Online Address: http://www.terratherm.com/default.htm.

14. Tetra Tech EM Inc. 2000. Draft Final Closeout Report. Naval Facility Centerville Beach, Ferndale, California. February.

15. Todd, Levi. Year. Publication or Report. Lessons Learned from the Application of In Situ Thermal Destruction of Hexachlorocyclopentadiene Waste at the Rocky Mountain Arsenal. Month.

16. U.S. Environmental Protection Agency. 2004. In Situ Thermal Treatment of Chlorinated Solvents Fundamentals and Field Applications. EPA 542-R-04-010. March.

17. Yargeau, Tedd, California EPA, DTSC. 2004. Email to Chitranjan Christian, Tetra Tech EM Inc., Regarding Questions on ISTD implementation at Alhambra. December 22.

18. Yargeau, Tedd, California EPA, DTSC. 2004. Telephone Conversation with Chitranjan Christian, Tetra Tech EM Inc. Response to Questions on ISTD implementation at Alhambra. November 2.

# APPENDIX E

# Additional Technologies Identified but Not Commercially Available

This Appendix presents technologies that were identified in the first edition (2005) of this report that are not currently commercially available.

## E-1    Xenorem™

Xenorem™ is an ex situ bioremediation technology that has been used to treat low-strength wastes containing chlordane, DDT, dieldrin, and toxaphene contamination. Xenorem™ uses an enhanced composting technology consisting of aerobic and anaerobic treatment cycles. Organic amendments such as manure and wood chips are added to contaminated soil, which can increase the final amended soil volume by as much as 40 percent (Ref. 37).

A self-propelled SCAT windrow incorporates the amendments into the soil and provides aeration to create aerobic conditions. High levels of available nutrients from the amendment increase the metabolic activity in the amended soil and deplete the oxygen content, creating anaerobic conditions. The anaerobic conditions promote dechlorination of organochlorine compounds. The length of the anaerobic phase is determined by bench-scale studies. At the end of the

| |
|---|
| **TECHNOLOGY TYPE: BIODEGRADATION** |
| **POPS TREATED: CHLORDANE, DDT, DIELDRIN, AND TOXAPHENE** |
| **MEDIUM: SOIL** |
| **PRETREATMENT: NONE** |
| **COSTS: $132 PER CUBIC YARD (COST IN 2000 USD)** |
| *FULL SCALE*<br>*EX SITU* |

anaerobic phase, the SCAT unit is used to mix the amended soil, creating aerobic conditions again. The anaerobic and aerobic cycles are repeated until the desired contaminant reductions are achieved. Typically, by the end of 14 weeks of treatment the organic amendments are spent. Soil samples are collected from the treated soil, and if the contaminant concentrations do not meet the cleanup goals, more organic amendments are added; the treatment is continued as long as necessary.

This technology was applied for a full-scale cleanup at the Stauffer Management Company Superfund site in Tampa, Florida. The site is a former pesticide manufacturing and distribution facility that operated from 1951 to 1986 (Ref. 20). Soil on the 40-acre site was contaminated with chlordane, DDD, DDE, DDT, dieldrin, molinate, and toxaphene. The Xenorem™ technology was applied to two 4,000-cy batches of soil. The first batch was completed in 2001 and the second batch was completed in 2002. The contaminated soil was excavated; screened; mixed; and amended with dairy cow manure, chicken litter, and wood chips. The amended soil matrix was then placed in a compost windrow. The temperature, oxidation-reduction potential, and moisture level of the amended soil matrix were continuously monitored (Ref. 20). Table 3-13 presents the performance data for both batches. Batch 1 was treated for a total of 24 weeks and achieved the site cleanup goals for chlordane, DDD, DDE, dieldrin, and molinate. After 12 weeks of treatment, Batch 2 achieved the site cleanup goals for chlordane, DDE, dieldrin, and molinate. The treatment of Batch 2 extended beyond 12 weeks; the final performance data for Batch 2 are not yet available from the vendor. Neither batch achieved the site cleanup goals for DDT and toxaphene. Typical treatment costs in USD using Xenorem™ were provided by the vendor and are approximately $132 per cy of contaminated soil (Ref. 22).

The Xenorem™ technology also was applied to a third batch of contaminated site soil at the site. Batch 3 was treated for one year but did not achieve the cleanup goals for chlordane, DDT, dieldrin, and toxaphene. Because the selected remedy did not fully meet the cleanup goals, the remedial design for the site is being modified. US EPA is awaiting details of the modification proposal. US EPA will prepare an Explanation of Significant Difference (ESD) fact sheet explaining the selection of a new remedy (Ref. 36).

### Table E-1. Performance of Xenorem™ Technology at the Tampa Site

| Pesticide | Site Cleanup Goal (mg/kg) | Batch 1 [a] | | | Batch 2 [b] | | |
|---|---|---|---|---|---|---|---|
| | | Untreated Concentration (mg/kg) | Treated Concentration (mg/kg) | Percent Reduction | Untreated Concentration (mg/kg) | Treated Concentration (mg/kg) | Percent Reduction |
| Chlordane | 2.3 | 3.8 | < MDL | NA | 4.5 | 1.2 | 75% |
| DDD | 12.6 | 26 | 9.3 | 65% | 24 | 14 | 42% |
| DDE | 8.91 | 6.6 | 2.1 | 68% | 6.1 | 2.6 | 57% |
| DDT | 8.91 | 82 | 9.8 | 88% | 196 | 14 | 93% |
| Dieldrin | 0.19 | 2.4 | <MDL | NA | 2.7 | 0.7 | 74% |
| Molinate | 0.74 | 0.2 | <MDL | NA | 0.4 | <MDL | NA |
| Toxaphene | 2.75 | 129 | 7.8 | 94% | 139 | 23 | 83% |

Source: Ref. 37

Notes:
MDL = Method detection limit (the MDL was not provided in the source document)
mg/kg = Milligram per kilogram
NA = Not available
[a] For Batch 1, treated concentrations are at the end of a 24-week period.
[b] For Batch 2, treated concentrations are at the end of a 12-week period.
Quantity treated: 4,000-cy of soil (Batch 1 and Batch 2).

Xenorem™ is a biodegradation process that uses an enhanced composting technology to treat various POPs in contaminated soil. The Xenorem™ process includes stages of aerobic and anaerobic treatments. Based on structural similarity of DDT, chlordane, toxaphene and dieldrin to other POPs described in section 2.6 of this report, this technology can potentially be used to treat other POPs. However, because of the specificity of biochemical reactions, this technology may or may not be effective in treating similar POPs. The last technology application occurred in 2002 at the Stauffer Superfund Site in Florida. Xenorem™ is a patented technology developed by Stauffer Management Company, a subsidiary of AstraZeneca Group PLC in Mississauga, Ontario, Canada. Recently, this technology was sold to the University of Delaware (Ref. 36). Additional information on the technology can be obtained from the Technology Transfer Corporation at the University of Delaware in Newark, Delaware. No fact sheet for this technology is currently available. Vendor contact information is provided in Section 5.0.

### E-2    Supercritical Water Oxidation

Supercritical water oxidation (SCWO) is an ex situ technology that has been used to treat solid and liquid wastes. It is potentially applicable to both low- and high-strength wastes containing POP contamination. SCWO occurs in an enclosed system at a temperature

> THE FACT SHEET PREPARED BY IHPA IS AVAILABLE AT
> HTTP://WWW.IHPA.INFO/RESOURCES/LIBRARY/

and pressure above the critical point of water (374°C and 22.1 x 10⁶ Pascal). Under these conditions, the gas-liquid phase boundary ceases to exist, and water is supercritical (that is, present in a fluid state that is neither liquid nor gas). Organic compounds have a higher solubility in supercritical water. An added oxidant such as oxygen or hydrogen peroxide reacts with dissolved organic contaminants in the supercritical water to form carbon dioxide, water, inorganic acids, and salts (Refs. 23 and 46).

The specifics of SCWO system design and operation vary. In general, currently available SCWO systems operate continuously, use corrosion-resistant materials in their reactors and process only fluid influents. One system marketed by Turbosystems Engineering Inc. blends a contaminated aqueous stream with an oxidant from a storage tank. The blended stream is pressurized, preheated, and passed into the SCWO reactor. Contaminants are destroyed inside the reactor, and the effluent is cooled, depressurized, separated into liquid and gas streams and discharged. SCWO technology is also available from General Atomics' Advanced Process Systems Division (Ref. 31).

> TECHNOLOGY TYPE: THERMAL-
> CHEMICAL DEGRADATION
>
> POPs TREATED: CHLORDANE, DDT,
> PCBs, DIOXINS AND FURANS
>
> PRETREATMENT: EXTRACTION/
> GRINDING, DILUTION
>
> MEDIUM: SOLID AND LIQUID WASTES
>
> *PILOT SCALE*
> *EX SITU*

The Assembled Chemical Weapons Assessment (ACWA) Program was established by the Department of Defense (DoD) in 1997 to test and demonstrate at least two alternative technologies to the baseline incineration process for the demilitarization of assembled chemical weapons (Ref. 6). In 2003, the Bechtel Parsons Blue Grass Team was awarded a contract to design, construct, test, operate, and close the Blue Grass Army Depot Destruction Pilot Plant using SCWO. The SCWO system is currently being constructed. SCWO was also selected for use at the Newport Army Depot to destroy 1,269 tons of liquid agent VX. Existing SCWO systems are limited to treating liquids and solids with a particle size of less than 200 microns suspended in a liquid. The process is best suited to wastes with less than 20 percent organic content (Ref. 44). SCWO treatment of solid wastes after they have been ground into a fine slurry has been demonstrated using feed materials containing up to 25 percent suspended solids (Refs. 6, 35, 44, 46 and 71).

Information regarding the SCWO technology is available from General Atomics' Advanced Process Systems Division (General Atomics) (Refs. 31 and 46). The SCWO process developed by General Atomics was selected for use as an ACWA technology to treat non-POPs such as GB, VX, H, HD, and TNT. However, no further information regarding process details, performance data, or costs could be obtained directly from General Atomics. Turbosystems Engineering Inc. also designs and markets SCWO systems in the US (Ref. 67). Turbosystems Engineering Inc. claims that its system can treat DDT and HCB; however, no performance data substantiating this claim are available in the information sources identified and used to prepare this report.

Supercritical Water Oxidation is a thermal-chemical degradation process that uses both, high temperatures and chemical additions to treat POP contaminated material. This technology has been proven to treat certain pesticides and industrial chemicals that are listed as POPs. Due to the high temperature requirement and the specific reactions under the treatment conditions of the technology, other POPs could also be potentially treated using Supercritical Water Oxidation. A commercial SCWO system developed by SRI International, USA and licensed to Mitsubishi Heavy Industries has been operational in Japan since 2002. The system is treating PCBs and uses sodium carbonate as an oxidant, which allows operation at a moderate temperature (380-420 °C) and mitigates potential corrosion problems. The most recent application of Supercritical Water Oxidation occurred in 2003. Currently, no further information regarding process details, performance data, or costs could be obtained directly from the technology vendor. Further technology information can be obtained by contacting the vendor using the information provided in Section 5.0.

## E-3    Vacuum Heating Decomposition

Vacuum heating decomposition is an ex situ technology for treating POP contaminated soil. This treatment is based on a technology used to remove zinc from zinc-plated steel by vacuum heating. Pretreatment is not required for POPs prior to decomposition by vacuum heating. The wastes are heated under vacuum conditions, where POPs are decomposed by pyrolysis and dechlorination reactions. The heating is regulated so the pressure ranges from 0.5 to 2,000 Pascals (Pa). Gaseous emissions pass through activated carbon prior to discharge.

POPs-related pesticides have been treated at one commercial site in Japan since 2004; however, no information is available for any full-scale projects treating POPs in the information identified and used to prepare this report (Ref. 44).

Vacuum Heating Decomposition technology uses high temperatures for thermal degradation under regulated pressure conditions to treat POPs such as; HCBs, PCBs and dioxins and furans. Due to the high temperature and pressure requirement of this technology other POPs could also be potentially treated using Vacuum Heating Decomposition. The vendor of this technology is Hoei-Shokai Co., Ltd. of Japan. This technology is not commercially available in the US. Currently, no further information regarding process details, performance data, or costs could be obtained directly from the technology vendor. Technology information can be obtained by contacting the vendor using the information provided in Section 5.0.

> **TECHNOLOGY TYPE: THERMAL-PHYSICAL DEGRADATION**
>
> **POPs TREATED: CHLORDANE, ALDRIN, DIELDRIN, ENDRIN, HCB, PCBS, DIOXINS AND FURANS**
>
> **MEDIUM: SOIL**
>
> **PRETREATMENT: NONE**
>
> *PILOT SCALE*
> *EX SITU*

> **THE FACT SHEET PREPARED BY IHPA IS AVAILABLE AT**
> HTTP://WWW.IHPA.INFO/RESOURCES/LIBRARY/

## E-4    CerOx™

CerOx™ is an ex situ electrochemical reaction technology that has been used in pilot tests to treat low-strength liquids containing POP contamination. CerOx™ uses cerium in its highest valence state (IV) to oxidize organic compounds, including POPs, to form carbon dioxide, water, and inorganic acid gases. The technology uses an electrochemical cell to produce cerium (IV) from cerium (III). Prior to treatment, solid waste such as soil or sediment is mixed with water to produce a fluid waste stream. This waste stream is injected with cerium (IV) from the electrochemical cell, agitated through sonication, and

> **POPs TREATED: CHLORDANE, DIOXINS, AND PCBS**
>
> **PRETREATMENT: SOIL AND SEDIMENT ARE MIXED WITH WATER TO PRODUCE A FLUID INFLUENT**
>
> **MEDIUM: LIQUIDS**
>
> *PILOT SCALE*
> *EX SITU*

CerOx™ treatment system,
Source: Ref. 15

transferred to a liquid-phase reactor. The liquid-phase reaction takes place at a temperature between 90 and 95°C and results in the destruction of organic compounds in the waste stream. During this process, cerium (IV) is reduced to cerium (III). Cerium (III) and unreacted cerium (IV) are returned to the electrochemical cell for recycling, and the treated medium is removed from the system. Gases produced during the liquid-phase reaction usually include carbon dioxide, chlorine gas, and unreacted volatile organic compounds (VOCs). These gases are processed through a gaseous-phase reactor that uses cerium (IV) to destroy VOCs. The remaining gases are passed through a scrubber to remove acid gases and are then vented to the atmosphere. Liquid from the scrubber is discharged (Ref. 15).

The information sources used to prepare this report did not describe any applications of CerOx™ systems at a pilot or full scale for treatment of POP-contaminated soil or sediment. CerOx™ systems have been used to treat POP-contaminated liquids. The first CerOx™ system was installed at the University of Nevada at Reno (UNR) to destroy surplus chlorinated pesticides and herbicides from the university's agricultural departments. Prior to use of this system by UNR, CerOx Corporation conducted proof of performance tests in May 2000. The medium treated was a pesticide-water emulsion. In one test, 71 percent by mass chlordane was mixed with water and fed to the system. The system is reported to have achieved a chlordane destruction efficiency of 99.995 percent in the gaseous-phase reactor (Ref. 5). Chlordane concentrations in the liquid effluent were not reported.

> **THE FACT SHEET PREPARED BY IHPA IS AVAILABLE AT**
> HTTP://WWW.IHPA.INFO/RESOURCES/LIBRARY/

The vendor later performed additional tests of the UNR system to determine the ability of CerOx™ to treat PCBs and dioxins (Ref. 71). A treatment test was performed on August 29, 2000, using a feed stream consisting of three commercially available dioxins dissolved in isopropyl alcohol. The dioxins in the feed stream were present at a concentration of 5 parts per billion (ppb). Two of three samples collected from the system's effluent contained dioxins at concentrations lower than their detection limit of 0.397 part per trillion (ppt). One sample had a dioxin concentration of 0.432 ppt. The UNR system was tested again on August 30, 2000, using a liquid sample from a remedial operation being performed in Fayetteville, North Carolina. The sample consisted of an isopropyl alcohol solution containing about 2 parts per million (ppm) PCBs. The system effluent contained PCB concentrations less than the minimum detection limit of 0.4 ppb PCBs (Ref. 15).

The technology was developed by CerOx™ Corporation in Santa Maria, California. CerOx™ Corporation offers a variety of CerOx™ treatment systems for commercial use. The systems range in size from modules with 25-gallon per day (gpd) treatment capacities to multimodular plants with 100,000-gpd treatment capacities (Ref. 15). This technology was included in the 2005 report; however, no information about this vendor could be found for this report.

www.ingramcontent.com/pod-product-compliance
Lightning Source LLC
Chambersburg PA
CBHW080642180526
45168CB00008B/3271